Praise for *The Healing Self*

"The wellness movement takes a quantum leap thanks to *The Healing Self.* It explains how the choices we make today, including choices such as diet and stress relief that affect our microbiomes and immune system, are crucial to lifelong health. No one explains this better and in greater depth than Chopra and Tanzi. Their combination of medicine and wisdom is unique."

—Rob Knight, Ph.D., Professor of Pediatrics and Computer Science and Engineering at the University of California (San Diego) and author of *Dirt Is Good: The Advantage of Germs for Your Child's Developing Immune System* and *Follow Your Gut: The Enormous Impact of Tiny Microbes*

"Deepak Chopra and Rudy Tanzi once again effectively harness their scientific training and passionate commitment to advancing well-being in this accessible, life-changing compendium of cutting-edge tools that will extend your health span and improve your sense of meaning, connection, and flourishing. As inspiring as it is informative, *The Healing Self* shows us how to use our minds to open our awareness and create daily routines to improve our physical and mental health. Bravo to our guides and to you for taking these practical steps toward healing in your life!"

—Daniel J. Siegel, M.D., Clinical Professor, University of California (Los Angeles) School of Medicine and New York Times bestselling author of *Mind: A Journey to the Heart of Being Human* and *Aware: The Science and Practice of Presence*

"The body has an extraordinary healing system. Yet most of us do not know how to activate that system to create vibrant health and prevent disease. Drs. Chopra and Tanzi map out a comprehensive strategy to turn on that healing system and engage your immune system in the healing response. In *The Healing Self* you will find a practical set of tools informed by the latest science for lifelong health and healing. This book has the power to change your health forever."

—Mark Hyman, M.D., Director of the Cleveland Clinic Center for Functional Medicine and the #1 *New York Times* bestselling author of *Eat Fat, Get Thin*

"As an academic dean and steward of population health, I understand the damaging effect of prolonged stress and inflammation on overall well-being. Research has shown us that many chronic diseases—such as diabetes and heart disease—can be greatly improved or even prevented through behavioral interventions. Deepak Chopra and Rudolph Tanzi's book *The Healing Self* provides a potential blueprint for lasting behavioral change through diet, exercise, rest, and positive human connection. This book has the power to enhance individual health and impact many of today's public health challenges."

—**Bess H. Marcus, Ph.D., Professor of the Department of Behavioral and Social Sciences and Dean of Brown University School of Public Health**

"In this landmark book, Drs. Chopra and Tanzi show us that true healing begins with self-healing, a journey of recognizing and nurturing the healing self and the acceptance of our 'dual role' as both the healer and the healed. The healing self is our very own nature, that intelligence which creates, nurtures, and sustains all life as we know it."

—**Paul Mills, Ph.D., Professor of Family Medicine and Public Health Chief of University of California (San Diego), Behavioral Medicine Division**

"With clarity and eloquence, Drs. Chopra and Tanzi masterfully marshal the latest scientific evidence demonstrating that the root cause of many illnesses and disease is chronic stress and inflammation. More importantly, *The Healing Self* gives us a clear and concise roadmap that allows us to use this information to heal ourselves and stay healthy for life."

—**James R. Doty, M.D., Professor of Neurosurgery, Director of the Center for Compassion and Altruism Research and Education (CCARE), and *New York Times* bestselling author of *Into the Magic Shop: A Neurosurgeon's Quest to Discover the Mysteries of the Brain and the Secrets of the Heart***

"The Healing Self is a quantum leap forward in the integration of science, medicine, wisdom, and health. As Deepak Chopra and Rudolph Tanzi powerfully show, our first line of healing lies inside us, and the choices we make today are crucial to lifelong wellness."

—**Arianna Huffington, founder and CEO of Thrive Global**

"*The Healing Self,* as revealed by Chopra and Tanzi, is a natural extension of the immunity and inflammation that can be managed by the concrete approach of these seasoned authors."

—Mehmet Oz, M.D., Professor of Surgery, New York–Presbyterian/ Columbia University

"Stress unequivocally affects the immune system and susceptibility to diseases. In *The Healing Self,* Chopra and Tanzi provide both a holistic and scientific perspective to understand this complex relationship and promote health."

—Eric J. Topol, M.D., Professor of Molecular Medicine, Scripps Research Institute, and author of *The Patient Will See You Now*

"Perhaps the major medical breakthrough of our time is the discovery that most, if not all, chronic disorders begin years earlier than the first symptoms. We've learned that about Alzheimer's—that it's twenty years in the making. *The Healing Self* reminds us how important it is that we care for our brains and our bodies at the earliest stage possible."

—Maria Shriver, journalist and founder of the Women's Alzheimer's Movement

"Most chronic disorders begin years earlier than the first symptoms. *The Healing Self* is a great resource in showing readers how to use this knowledge for enhancing immunity, slowing the aging process, and going beyond conventional prevention."

—Dean Ornish, M.D., founder and president, Preventive Medicine Research Institute; Clininal Professor of Medicine, University of California (San Francisco); and author of *The Spectrum*

Deepak Chopra, M.D., and
Rudolph E. Tanzi, Ph.D.

The Healing Self

A Revolutionary
New Plan to Supercharge
Your Immunity
and Stay Well for Life

HARMONY
BOOKS · NEW YORK

Published in the United States by Harmony Books, an imprint of the Crown
Publishing Group, a division of Penguin Random House LLC, New York.
crownpublishing.com

Harmony Books is a registered trademark, and the Circle colophon is a
trademark of Penguin Random House LLC.

Library of Congress Cataloging-in-Publication Data
is available upon request

Hardcover ISBN 978-0-451-49552-5
Ebook ISBN 978-0-451-49553-2
International Edition 978-0-525-57433-0

Printed in the United States of America

Illustration on page 15 by Digital Mapping Specialists
(original illustration source courtesy of Blake Gurfein)
Jacket design by Pete Garceau

10 9 8 7

First Hardcover Edition

To the healer in everyone

Contents

Wellness Now—Many Threats, One Great Hope

At the end of July 2017, a startling medical story came across television and the Internet. It was a tip-of-the-iceberg story, although few people realized it at the time. There was too much background noise from the usual stream of health risks people were supposed to heed. Among the latest risks: Working more than fifty-five hours a week can be bad for your health. Pregnant women are at higher risk of not getting enough iodine.

These were not tip-of-the-iceberg stories—more like the drone of familiar advice that most people have learned to shrug off. But one item was different. Twenty-four experts on old-age dementia—the greatest health threat around the world—were asked to assess the overall chances for preventing every kind of dementia, including Alzheimer's disease. Their conclusion, published in the prestigious British medical journal *The Lancet*: One-third of dementia cases can be prevented. There is currently no drug treatment to cure or prevent dementia, so this was startling news on the face of it.

What was the key to preventing dementia? Lifestyle changes, with a different focus at every stage of life. The experts singled out nine specific factors that accounted for around 35 percent of dementia cases: "To reduce the risk, factors that make a difference include getting an education (staying in school until over the age of fifteen); reducing high blood pressure, obesity, and diabetes; avoiding or treating hearing loss in mid-life; not smoking; getting physical exercise; and reducing depression and social isolation later in life."

One item from the list was startling: staying in school until at least the age of fifteen. What in the world? A dreaded condition of old age could be reduced by doing something when you are a teenager? For that matter, it was also a little peculiar that addressing hearing loss in middle age was related to a lower risk of dementia. Something new was going on. If you looked close enough, this news story was signaling a trend in medicine that promises to be a major revolution.

Not just in dementia, but across the board researchers are drastically pushing back the timeline of disease and life-threatening disorders like hypertension, heart disease, cancer, diabetes, and even mental disorders like depression and schizophrenia. When you catch a winter cold, you notice the symptoms and realize, with annoyance, that you were exposed to the cold virus a few days earlier. The incubation period was short and invisible; only the appearance of symptoms told the tale. But lifestyle disorders aren't like that. Their incubation period is invisible but very long—years and decades. This simple fact has become more and more critical in medical thinking. Now it looms larger perhaps than any other factor in who gets sick and who stays well.

Instead of focusing on lifestyle disorders when symptoms appear, or advising prevention when high risk has developed, doctors are probing into normal, healthy life twenty to thirty years earlier. A new vision of disease has been emerging, telling us some very good news. If you practice lifelong wellness, beginning as early as childhood, the many threats that attack us from middle age onward can be defeated—the secret is to act before any sign of threat appears.

This is known as "incremental medicine"—the iceberg of which a single story about dementia is the tip. Take the seemingly strange finding about education. Experts estimate that dementia could be reduced by 8 percent globally if kids stayed in school until they were fifteen, one of the biggest single reductions on the list. The reason why traces a long trail. The more educated you are, the more information your brain stores and the better it accesses what you've learned. This buildup of information, starting in childhood, leads to something neuroscientists have identified as "cognitive reserve," a boost to the brain in terms of added connections and pathways between neurons. When you have this boost, the memory loss associated with Alzheimer's and other forms of dementia is countered, because the brain has extra paths to follow if some grow weak or diseased. (We discuss this in more detail in our section on Alzheimer's at the end of the book.)

As medical logic goes, long trails are changing everyone's thinking, because they exist in many if not most diseases. Suddenly it's not about isolated factors like not smoking, losing weight, going to the gym, and worrying about stress. It's about a continuous style of living where self-care matters every day in every way. Not smoking, losing weight, and going to the gym still have their benefits. But lifelong wellness isn't the same as lowering your risks for disorder A or B. Only a holistic approach will ultimately work. Wellness is no longer just a valid alternative to regular prevention. It's the iceberg, the four-hundred-pound gorilla, and the elephant in the room rolled into one. Wellness is the great hope springing up all around us. When the public gains full knowledge of this fact, prevention will never be the same. But to grasp how radically things will change, we have to back away and examine the current situation in health care, where threat increasingly overwhelms hope.

The Immunity Crisis

Modern medicine makes so many headlines every day that they blur together, and it becomes nearly impossible to sort out what's important here and now. It can seem like just being alive is a risk to your health. So let's simplify things. The most urgent crisis facing human health today comes from something most people take for granted: their immunity. This is the crunch where health and disease clash. Immunity is medically defined as the defense your body mounts against invasive threats, medically known as pathogens. In common parlance these are vaguely lumped together as germs, the host of bacteria and viruses that exists for one purpose, not to make us sick but to promote their DNA. As a biosphere, the Earth is a vast arena in which DNA evolves, and although we feel special, even unique, as human beings, our DNA is only one gene pool among millions.

Immunity is what keeps our genes ahead of survival threats, and it has been brilliantly successful to date. Despite catastrophic events in the history of disease that swamped our DNA like a tsunami—smallpox in the ancient world, bubonic plague in the Middle Ages, AIDS in modern times, just to mention a few terrible examples—our immune system has never faced the level of threat it faces today. Smallpox, plague, and AIDS didn't annihilate *Homo sapiens* as a species, nor has any other pathogen, because three factors saved us:

1. None of these diseases is so communicable that every person on Earth could catch it. Either the germ couldn't survive in the open air or people lived far enough apart that the disease couldn't survive while crossing the distance between them.

2. Our immune system is capable of improvising new kinds of genetic response very quickly, through a process known as hypermutation. This constitutes an immediate tactic for combating unknown pathogens the moment they enter the body.

3. The rise of modern medicine has come to the rescue with drug and surgical treatments when the body's immune system can't fight a disease on its own.

These three powerful agents are all necessary for you to stay healthy, but they may have reached a tipping point. The global competition among millions of strains of DNA has been heating up to alarming levels. Immunity can no longer be taken for granted no matter what part of the world you live in. Our overburdened defense system against disease is steadily crumbling. That's because of a host of problems that actually go beyond the scary potential for a new epidemic, whether from the Zika virus or avian flu. Those threats grab the headlines, but with far less publicity the whole health-care situation is fraught on multiple fronts.

Why a Tipping Point Is Coming Closer

- Modern travel has drastically reduced the distance between all peoples, making it much easier and faster for new pathogens to spread and find new hosts.

- Viruses and bacteria mutate faster than ever because new human hosts keep multiplying at unheard-of rates of population growth.

- New drugs cannot be developed as fast as potentially hazardous strains of DNA that mutate at the microscopic level of bacteria and viruses.

- While the threat keeps mounting, medical systems are burdened by inertia, income inequality, frightening expense, and massive scientific complexity.

- Prevention has existed for fifty years but has failed to eradicate persistent heart disease, hypertension (high blood pressure), type 2 diabetes, widespread depression and anxiety, and the latest epidemic, obesity.

- An aging population faces a higher incidence of cancer and the threat of dementia, chiefly through Alzheimer's disease.

- Older people have higher expectations, wanting to be healthy and active well past sixty-five or even eighty-five.

- Turning into a drug-dependent culture has caused a host of problems, including opiate addiction, and even when drastic problems are sidestepped, it's estimated that the average seventy-year-old takes seven prescription drugs.

- New strains of "superbugs" like MRSA are staying ahead of antibiotics and antivirus medications.

This list is too long and alarming to ignore. Your health is intertwined with every factor on it, and as serious as it would be if the world went past a tipping point, the immediate issue is keeping yourself from going past it.

The secret is to expand the definition of immunity and then to use a rich array of choices with one aim, to supercharge your immunity. According to the standard understanding, your immunity gets stronger basically when you develop a new antibody against this winter's flu virus, for example, but not when you eat an anti-inflammatory diet. Yet it is now recognized that low-grade chronic inflammation, a condition with almost no overt signs you would generally be able to detect, is linked to more and more disorders, including heart disease and cancer. In an expanded definition, fighting inflammation would be absolutely critical to total immunity.

Total Immunity and the Healing Self

Total immunity is the measure of holistic health. A crucial aspect was covered in our book *Super Genes*, where we introduced the concept of DNA as something dynamic, ever changing, and totally responsive to a person's lifetime of experience. If DNA were frozen, locked up, and unchanging, then supercharging your immunity would be wishful thinking. Such a viewpoint held sway, however, for decades. A new era began as soon as DNA was freed up through a model that showed how totally our gene activity is affected by the world around us. The competition among global strains of DNA suddenly became much more urgent.

We felt that total immunity demanded more. What about the mind and its effect on health? What about behavior, habits, and the contribution of the family? Why should germs be given more importance than other common causes of disease, for example cancer, which is almost always unrelated to invading micro-organisms? To encompass everything, it was necessary to abolish the boundary between mind and body. A leap of imagination was called for. Therefore, we're introducing a new term, *the healing self*, that satisfies the real meaning of wholeness. Two roles that are involved every day in keeping us healthy have long been kept separate. The first role is the healer; the second is the one being healed. These two roles are currently played by an outside healer and the patient who depends on him or her. The outside healer doesn't necessarily mean an MD. The important word here is *outside*, which puts the burden of care on someone besides you.

The traditional separation of roles isn't realistic as far as your body is concerned. Immunity is centered on the self. A doctor's role isn't to boost your immune response from day to day. Medical care becomes active, for the most part, only when symptoms appear, and by then the immune response has broken down. In the broader picture, the entire healing response has broken down, of which immunity is the centerpiece. There has always been a mismatch between what medicine can do and what the body needs if it wants to protect itself in the global competition of DNA.

The doctor-patient partnership isn't designed for meeting the competition and winning. But the healing self, by merging healer and healed, can surmount the looming threat. (Important note: We certainly aren't advising you to ignore or avoid a physician's care when it is needed.) If you become proactive about your own immunity, the whole situation changes. Looking back at the list of threats we began with, some urgently needed improvements can occur once you learn what it means to adopt the healing self.

Benefits of the Healing Self

- It is noninvasive and involves no reliance on external therapies.

- It maintains natural balance and boosts your immune system through lifestyle choices.

- Lifestyle choices can prevent many forms of cancer and hold promise for preventing Alzheimer's disease and even reversing symptoms of dementia.

- Successful aging will consist of a long healthspan as well as a long lifespan.

- Drug dependency is staved off because healing occurs before the stage of symptoms begins. The vast majority of drugs are prescribed late in the disease process, a stage you don't have to reach if you act early enough. This is true for almost every lifestyle disorder, including heart disease and cancer, disorders that create the strongest need for drug treatments.

These are practical outcomes from adopting the dual role—healer and healed—of the healing self. What makes it all possible is raising your awareness. What you aren't aware of you can't change. The biggest thing most people aren't aware of is the very possibility of self-healing. Let's see how this applies to immunity.

All living things need to repel outside threats to their DNA. Modern medicine recognizes two types of immunity, passive and active. As the term implies, *passive immunity* is beyond your control, being genetically based. You inherited your mother's antibodies in the womb, and after you were born other antibodies were transferred in her breast milk. (There are also medical means to pass on antibodies from another person through blood and plasma infusions or even the transfer of another person's T-cells, but these methods are rarely used and carry high risks.)

The other kind of immunity, *active immunity*, fights disease organisms (pathogens) directly on the front lines. All living creatures above a certain level have innate, or inborn, immune defenses, including plants, fungi, and multicelled animals. The *innate immune system* is very general. It can detect that a pathogen is invading the host and then release chemicals to fight back. But active immunity in higher animals, including humans, has evolved far beyond this stage. We have specific immune cells (for example, T-cells and B-cells) that have evolved to a nearly miraculous capacity for responding to invaders.

Myriad times a day the immune response identifies one kind of germ from thousands of possibilities and rushes into action to chemically disable the invader. Specific white cells engulf its remains, and these are quickly flushed out of your body. On the other hand, you can't help but notice when this precise sequence of events makes mistakes. The result is an allergy, which is the result of mistaking a harmless substance (pollen, cat dander, gluten, etc.) for an enemy, giving rise to a full-blown chemical reaction that is often harmful. This immune response can often be due to bacteria that ride along with the substance into the body. Even pollen has a microbiome! In other cases, the immune system may be activated to attack specific proteins in the body, causing an autoimmune disorder like rheumatoid arthritis or lupus.

Staying alive depends on minimizing such errors. Therefore, every disease your ancestors successfully fought off is stored as the antibodies you inherited, and when you ward off a new illness, like a new strain of flu, you add to this vast memory bank. Although the function of active immunity was discovered as far back as 1921 by the English immunologist Alexander Glenny, its precise mechanisms waited for decades to be understood. The picture is incredibly complex, biologically speaking, yet at least one external method for boosting active immunity is more than two centuries old: vaccination.

As we all learned in school, in the late 1700s the rural English physician Edward Jenner developed the first vaccine—and became known as "the father of immunology"—after he observed that milkmaids generally were immune to smallpox, a disease that had reached epidemic proportions. In France the philosopher Voltaire estimated that 60 percent of the populace contracted smallpox and 20 percent died of it. Jenner's insight was to take pus from a milkmaid who had contracted a much milder disease, cowpox, and inject it into his patients to transmit the same immunity that the milkmaid had.

Despite the current controversy surrounding vaccinations in some quarters, what Jenner established was proof that active immunity can be boosted. One does not have to wait until the course of evolution, which occurs over tens or hundreds of thousands of years, brings an improvement. The standard recommendations about diet, exercise, good sleep, and maintaining a good weight all benefit a person's immune status. These standard recommendations appear on Harvard Medical School's health website (www.health.harvard.edu), with two additions for avoiding infection: remembering to wash your hands frequently and cooking meats thoroughly.

Yet on the question of boosting the immune response itself, the Harvard website is skeptical:

> Many products on store shelves claim to boost or support immunity. But the concept of boosting immunity actually makes little sense scientifically. In fact, boosting the number of cells in your body—immune cells or others—is not necessarily a good thing. For example, athletes who engage in "blood doping"—pumping blood into their systems to boost their number of blood cells and enhance their performance—run the risk of strokes.

The Harvard Health Publishing website goes on to say: "But that doesn't mean the effects of lifestyle on the immune system aren't intriguing and shouldn't be studied. Researchers are exploring the effects of diet, exercise, age, psychological stress, and other factors on the immune system response, both in animals and in humans. In the meantime, general healthy-living strategies are a good way to start giving your immune system the upper hand."

The main reason for this skeptical attitude is that there are so many kinds of cells in the immune system that perform so many functions. But on the contrary side is powerful evidence from the mind-body connection. A variety of psychological states from grief to depression lower people's immunity, making them more susceptible to getting sick. This deterioration in immunity can't be seen under a microscope; it doesn't show up as physical changes in specific cells. There are not many studies that directly connect stress, for example, to physical changes in the immune system, yet the connection between high stress and getting sick has been well documented and is doubted by no one. If we expand our definition of immunity to everything that keeps us healthy, there is even more evidence about how lifestyle disorders like hypertension and heart disease become a greater threat when someone is poor, depressed, lonely, or living without social support.

These findings all point in the same direction. Immunity can be transformed into total immunity, but not by restricting our focus to the immune system, which includes only the physical side. The mind must be given equal importance, which is why *self* is the key word in the healing self.

The Mystery of Healing

Self sounds like something psychological, an invisible entity that you possess but that is unrelated to your body. If you develop an ovarian cyst or high blood pressure, those are problems rooted physically in the body, not the self. But is this really so? How you see yourself today makes a huge difference in what your body will be like tomorrow. Imagine that two strangers knock on your door. Both have surprising propositions.

The first stranger says, "I'm an MD, and I do advanced research on aging. My life's goal has been to find a pill that will alter the genes that cause aging. I think I've found a promising formula, and we need subjects to test it on."

He holds up a bottle of tiny blue pills.

"The trials start today, and I'd like you to volunteer," he says. "This is a blind trial. You'll take these pills twice a day for six months. Half of the subjects will be getting a dummy pill, a placebo. But just think what this could mean, the reversal of aging. Why should we accept that growing old is inevitable when we can unlock the genetic key that will change everything?"

His excitement impresses you, but the second stranger is wearing a faint smile. You ask her if she's part of the same drug trial.

"No, but I am here to show you how to reverse your age," she says. "No drugs or placebos are involved. Your age will start to reverse in around five days. After a week you can expect a lot of other beneficial changes. My experiment is short, but effective." She points to the first stranger. "His drug could have serious side effects. The FDA will have to approve his experimental drug if it shows results, and the approval process costs hundreds of millions of dollars and takes years to complete." The faint smile returns to her lips. "Of course, the choice is yours."

Which would you choose? Although we set up the situation as imaginary, in fact it's very real. Drug companies are constantly testing anti-aging drugs, with the most recent trend involving altering your DNA. There could be breakthroughs that will make a huge difference in human aging, long considered "a one-way street to incapacitation," to quote Professor Ellen Langer, a Harvard psychologist who has performed remarkable experiments of her own. Yet Langer could easily be the second stranger at your door. Professor Langer has a track record for reversing the signs of aging and extending longevity without drugs. In fact, she bypasses the body altogether and goes straight to the mind.

Langer's most famous experiment worked as follows. In 1981 eight men in their seventies, who were in good health but showing signs of age, were bused to a former monastery in New Hampshire. When they entered, the men found themselves immersed in the past, specifically the year 1959, listening to the crooning of Perry Como. They dressed in clothes suitable to that year. They watched a black-and-white TV and read newspapers filled with stories about Castro's takeover in Cuba and the hostile attitude of Nikita Khrushchev, premier of the Soviet Union. For a movie they watched Otto Preminger's *Anatomy of a Murder,* which came out in 1959, and sports talk focused on bygone figures like Mickey Mantle and Floyd Patterson.

As a control, another group of eight men lived as they normally would but were told to reminisce about the past. The time-capsule environment group were told something very different—they were to act exactly as if it were 1959 and they were twenty years younger. By any reasonable medical standards, the results of the pretend time travel should have been nil. But Langer had done earlier studies at Yale with elderly residents of nursing homes. She discovered that signs of aging, particularly memory loss, could be reversed through the simplest positive reinforcement.

Giving someone an incentive to remember, such as small rewards that depended on their test performance, brought back memory that everyone else had assumed was irreversible.

But even Langer didn't expect the dramatic results of her total-immersion experiment. Before entering the time-capsule environment, the men were tested on various markers of aging, such as grip strength, dexterity, and how well they could hear and see. At the end of the five days, the group that was immersed in the world of their younger selves showed improved flexibility, dexterity, and posture. They also improved on seven out of eight measures, including better vision, a startling finding. They looked younger as assessed by outside judges. These results were significantly better than in the control group, who also showed improvements in the same physical and mental areas through reminiscing about the past—for example, 63 percent of subjects in the time-capsule group scored higher on an intelligence test as compared with 44 percent of the control group.

"What matters here is what actually happened," Langer explains. "Men who changed their perspective changed their bodies." Thirty-six years ago Professor Langer was proceeding more or less intuitively. In 2017 we have research that indicates how changing experiences can alter gene expression and train the brain to continue developing new pathways, as we do when we learn new things or change our perspective (more on those breakthroughs in later chapters).

(In 2010, BBC One produced a TV series called *The Young Ones*, in which six aging celebrities lived together in a setting straight out of 1975. As in Langer's previous experiment nearly thirty years earlier, the participants seemed to get younger in front of our eyes. One celebrity who arrived barely able to bend over to put on his shoes found new suppleness on the dance floor. In general, everyone progressively began to look younger, from their posture to their facial expressions.)

The reversal of aging is very closely tied to healing, because both have long been considered totally physical and confined to bodily processes that proceed independent of the mind. Langer was among the first to explode these assumptions. It's easy to get lost in the fascination and mystery of why pretending to live in the past should change a person so quickly. But the most important clue is that the changes were holistic. Doctors are trained to deal with the body one organ, tissue, or even cell at a time. There is no medical rationale for how so many functions can improve at once, especially through playacting. Langer's results leave the placebo effect in the dust, because the placebo effect depends upon fooling a patient that he is taking a potent drug when all that is administered is a dummy pill.

In the time-travel experiment, no promises were made, no expectations raised. The only medicine involved was a new experience, and that was enough to confound all medical assumptions up to that time.

In one of her earlier experiments, Langer went into a retirement home and again divided her subjects into two groups. Both were given some houseplants for their room. One group was told that they were responsible for keeping the plants alive, and that they could make choices in their own daily schedule. The other group was told that the staff would tend the plants, and in addition they were given no choice in their fixed daily schedule. At the end of eighteen months, twice as many subjects in the first group were still alive compared with the second group.

The entire medical community should have had an "aha" experience when these experiments were performed. Decades later, using new experiences as a means of healing the aging and afflicted has become more feasible. Retirement home residents are given pets to take care of. Alzheimer's patients have been shown to improve while listening to music. In fact, Rudy and his colleagues have produced an app called SPARK Memories Radio to provide music therapy to Alzheimer's patients. A caregiver family member enters the birthdate of the patient and any information available about their tastes in music. The app then plays songs that were hits when the patient was between thirteen and twenty-five years old, since this is the music that people generally bond with emotionally for the rest of their lives.

E-mails from users poured into Rudy's team, recounting how early-stage Alzheimer's patients became calmer and less agitated, and how late-stage patients who were vegetative suddenly "woke up" again. One family told the story of their afflicted father, who was suffering in the late stages of the disease and had not spoken for months. After hearing five songs from his youth, he suddenly sat up in bed and started telling a story about a red pickup truck and his first girlfriend, providing perhaps too many details! The family was blushing with embarrassment, but they were thrilled to hear him speaking again, so happily and vibrantly. Similarly, one can find YouTube videos of Parkinson's patients who can barely walk without the support of a nurse suddenly finding their balance and even beginning to dance when music is played. This is the healing power of music or, more precisely, the healing power of our responses to pleasurable memories.

In short, we are entering a golden age for health and healing, largely depending on how each person employs the most common and yet most powerful tools at everyone's disposal: everyday experience, simple lifestyle choices, and techniques to increase awareness. The notion is actually ancient in origin. The medieval Indian philosopher and sage Adi Shankara declared that people grow old and die because they see other people grow old and die.

The Bodymind

Thirty years ago, doctors were suspicious of the mind-body connection, which aroused skepticism because, unlike the heart or a flu virus, the mind is invisible and nonphysical. Today, thanks to decades of research into how the brain communicates with every cell in the body, trying to find a bodily process that *isn't* influenced by the mind has become the real challenge. The brain, which was once the emperor of the mind, has been deposed. "Mind" is spread throughout your body. A heart or liver cell doesn't think in words and sentences, but it sends and receives complex chemical messages all the time. The bloodstream, along with the central nervous system, is an information superhighway teeming with traffic as 50 trillion cells contribute to a united goal: remaining alive, healthy, and thriving. Below is how the pathways of the information superhighway actually look.

To any medical student today or decades ago, the organs in this illustration are the familiar stuff of medical knowledge. But in the future, the added text will become just as standard. An educated physician will need to know everything about the "signaling pathways" that lead from the brain and back again. These pathways are actually what holds your body together. Unless each cell is directed what to do, kept informed about 50 trillion other cells, and plays its part in the body's holistic balance, there is no body, only a collection of detached independent cells, like the ones that make up a coral reef or a jellyfish.

Decades of research went into validating that the information superhighway is real, and even today more findings are proving how detrimental the separation of mind and body really is. In this book, we will

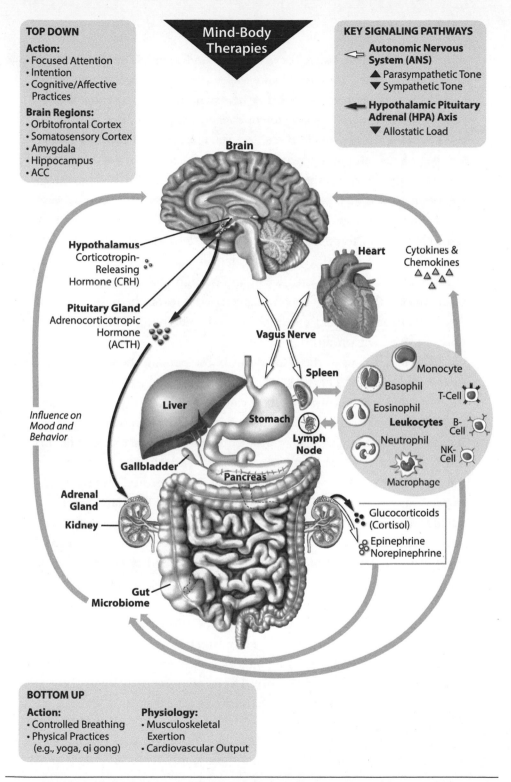

Mind-Body Therapies

drop the artificial division between body and mind. The proper term should be *bodymind*, for sound biological reasons. The same brain chemicals known as neurotransmitters—the essential molecules that enable your brain to function—are present everywhere, including your intestines. That discovery, made three decades ago, stunned medical science and helped fuel an intelligence explosion.

Suddenly the immune system, which is physically separate from the brain, was understood as part of a vast network of chemical messages throughout the body that rivals the messages sent by the brain— researchers began to refer to the immune system as a floating brain. It doesn't matter today that the mind-body connection is invisible, because at the molecular level it isn't. There are enough chemical clues to convince anyone that mood, beliefs, expectations, fears, memories, predispositions, habits, and old conditioning—all centered in the mind—are critical to a person's health.

Which brings us to the crux of this book. Among the processes that can be influenced by a person's awareness, healing is one of the most vital. Cells use their own form of chemical consciousness already. The immune response is awake and aware all the time, constantly monitoring itself, staying vigilant for any possible invader or other outside threat. The immune response is as self-sufficient as someone's heartbeat or breathing. Yet immunity as a built-in response, which every medical student learns as basic knowledge, has a gaping flaw in it. To find the flaw, pause and simply take a deep breath. There's the flaw, staring everyone in the face. Breathing is an automatic, involuntary function, but you can step in and make it voluntary anytime you want. The same ability extends almost everywhere. You can voluntarily induce the stress response by going to a horror movie. You can alter your metabolism by exercising or changing your diet. Get into a sexual situation and you bring big-time changes to all of the above, and more. The dividing line between what happens automatically and what happens voluntarily isn't fixed. Choices matter, and thus the healing self comes into play. On its own the body knows how to survive; it's up to us to teach it how to thrive.

PART ONE

The Healing Journey

1

Getting Real, Getting Started

Let's get real about staying healthy. Everyone wants to stay healthy as long as possible, but we're confused about how to do this. Conflicting information keeps appearing, backed up by studies that disagree as often as they agree. Eagerly followed fads come and go. Even very basic questions—Is milk good for adults? Do eggs increase cholesterol levels? How is obesity connected to type 2 diabetes? Why are allergies on the rise?—have been thrown into doubt.

We wind up taking the attitude that life is a gamble, and anyone who stays vital and vigorous for seventy or eighty years has been very lucky. The deeper reason we hold this attitude is that we feel the odds are stacked against us. Life isn't an upward arc. After your prime years, getting sick is inevitable. Every adult is statistically at risk for heart disease and cancer, the two leading causes of death in this country. Most people's greatest fear, Alzheimer's disease, apparently strikes at random and is incurable.

The gambling model for staying healthy is taught in medical school, only in a more scientific way. Despite all the marvels of modern medicine, a great deal remains uncertain. A specific cause of disease like a cold virus only makes a certain percentage of people sick, not everyone. Standard treatments all involve some degree of unpredictability, working better for some patients than others, and sometimes not at all. Reducing risks is how prevention is defined. By eating right, exercising regularly, and avoiding toxins like alcohol and tobacco, a person isn't actually attacking the cause of major disorders like diabetes, coronary artery disease, and cancer. Instead, the odds of getting sick are going up or down. The average person doesn't realize that these risks apply to big groups as measured by statistics. They don't predict what will happen to the individual. There will always be someone who does everything right but gets sick anyway, while someone else who has paid almost no attention to their health dodges the bullet.

Even if you're blessed with good luck, the day will come when the best doctors in the world cannot help you. Through no fault of your own, there will be a breakdown in your health, and the casino will start to gain its advantage. Here's why.

Seven Reasons Medical Care Stops Working

- The doctor doesn't know what caused you to get sick.
- There's no drug or surgery that will resolve the situation.
- The available treatments are too risky, toxic, expensive, or all three.
- The side effects of the treatment outweigh the benefits.
- Your condition is too far advanced to be reversed.
- You're too old to treat safely or with much hope of recovery.
- Somewhere along the line, a doctor made a mistake.

When any of these breakdowns in medical care occur, whatever happens next is out of your control, and your doctor's. After three centuries of scientific medicine making huge strides—a legacy the authors deeply respect—it's becoming obvious that the gambling model for staying healthy needs to be replaced. Too many unacceptable things are happening:

- People are living longer and yet on average suffer eight to ten years of bad health and one to three years of disability at the end.
- Cancer is still approached with grim fatalism despite the fact that up to two-thirds of cancers are preventable.
- An estimated 400,000 people die every year due to medical mistakes.

- The average person feels helpless, confused, and anxious about getting sick and going to the doctor.

These unacceptable things arise when the gambling model takes hold and you throw the dice with your future. The most unacceptable thing of all is losing control. People dread the notion of falling into the hands of doctors and winding up in the hospital. But there is an alternative. The healing self is the choice maker who steps into the arena of everyday life and steers mind and body toward a *lasting* healing response. A paper cut goes away after a day or two; last winter's cold is a distant memory. The healing self, on the other hand, is long range. You set out to become whole, which is the only viable strategy for remaining healthy over a lifetime.

It's amazing how far the human body has evolved to make healing possible. You now have an opportunity to evolve *consciously*, making choices that will radically upgrade your immunity to disease, slow down and reverse the aging process, and boost the healing response. These goals aren't achievable by gambling, but they can be achieved when you adopt a new model, the healing self.

In the new model, everything comes down to the process shown in the following diagram:

Disruption ⟶ Healing response ⟶ Outcome

Disruption = Any health threat: an invading virus or bacteria, a physical wound, a stressful event, distortions at the cellular or genetic level, mental distress, and the like

Healing response = A reaction to the disruption that restores balance in either mind or body

Outcome = A return to the normal, undisrupted state of balance

As you can see, the terminology is very general. Any experience can be a disruption, not necessarily a bacteria or virus. The memory of a past trauma can massively disrupt the body, as can losing your job or simply giving in to the impulse to have a double cheeseburger with fries. Likewise, the body's response to a disruption involves the entire messaging system of the information superhighway. Whatever returns the body to a normal state of balance counts as healing.

This approach is gaining traction in contemporary medicine as the *whole-system approach*, about which we will have much to say. *Whole system* is simply another way to say *bodymind*. It looks beyond the artificial medical-school divisions into separate organs and the old skepticism about the mind-body connection. When a happy event occurs, such as falling in love, the whole system responds as messages course through the bloodstream, central nervous system, and immune system. When a tragic event occurs like losing a loved one, the response is just as holistic, but the combination of chemicals in the signaling process is very different. What you experience subjectively as love or grief must have a precise configuration in the bodymind. If that didn't exist, you wouldn't have the experience.

The whole-system approach isn't just a bright shiny new model to replace the old ones—it comes closer to reality. Nature doesn't recognize human-made categories. Body and mind are one domain, and every organ, tissue, and cell works toward the same goal: sustaining life. Yet the sober truth is that our bodies haven't evolved fast enough to cope with the disruptions we're forcing on them. The whole-system approach reveals holistic problems as well as holistic solutions. Consider the current epidemic of obesity facing all age levels in America. Just one factor—excessive sugar intake—is a major contributor to obesity, type 2 diabetes, and least suspected of all, heart disease. You can eat sugar today and notice no signs of these creeping disorders, but your pancreas knows that the demand for insulin is too high; your digestive system knows that too many useless calories are being converted to fat; your hypothalamus knows that the quick energy of a sugar high throws your metabolism out of balance.

Powerful as the innate healing response is, it depends on evolution before a major shift can occur, which is far too slow. The only viable strategy is to intervene with conscious choices that the bodymind can absorb and adapt to. A double cheeseburger with fries is known to cause inflammatory markers to appear in blood plasma (the straw-colored liquid that is left in your blood after the solid part, chiefly red blood corpuscles, is removed), along with floating particles of fat. This happens within a few minutes and lasts upward of six hours. During that time, your body is experiencing a disruption. In response, your liver will rev into gear to process the excessive load of fat, and your immune system will attempt to combat the surge of inflammation. The immediate outcome is likely to be very undramatic and seemingly innocuous. But the drip, drip, drip of such disruptions has long-range damaging effects.

If you live your life unconscious of what's happening to the whole system, you are adhering to the gambling model of health. If you become aware of the downside of a double cheeseburger with fries, you might swear off such an indulgence, and your body will thank you for it. But temptation is constant and giving in takes only a minute, not simply with a cheeseburger but with all kinds of fatty, salty, overly sweet, processed, and junk food.

The only way to get real is to make a major shift into a healing lifestyle, one that isn't chopped up into small temporary choices—even very healthy ones—but rises to the level where the whole system is cared for.

What Can the Healing Self Do?

Imagine for a moment two patients, A and B, who feel feverish and go to the doctor. Patient A encounters a full waiting room and is told that the doctor is running late by thirty minutes. In reality, the wait is over an hour, and when A gets to see the doctor, she's feeling a bit tense and put out. In a businesslike way, the doctor takes her temperature, does a cursory exam, and writes a prescription for antibiotics.

"You might have a low-level infection," he says. "Let's see how this works for you. If you're catching a cold or the flu, your fever will get worse and then get better. See you in two weeks. The nurse out front can make an appointment."

This scenario is fairly typical of everyday visits to primary-care physicians, and each of us knows the routine. Nothing A's doctor told her is untrue or outside normal practice—she got routine care.

Patient B finds an empty waiting room and sees the doctor immediately. He asks about her fever and wants details about when it started and how badly it might have affected her sleep, her mood, her energy level, and her appetite. He investigates to see if B has had similar fevers in the past and, if so, how they were resolved. Did they go away on their own or were medications necessary? This interaction takes more than a few minutes, but the doctor looks interested and never impatient. Patient B finds his manner reassuring.

"Most of the time this kind of low-grade fever is the symptom of a cold or the flu," the doctors says. "Over the next few days, call me whenever you feel a need. Once we monitor what's going on, we'll have a better idea of what to do."

The second doctor sounds ideal, but there's one hitch: he's a fantasy. Few if any patients receive the kind of unhurried, sympathetic attention that our fictitious Patient B encountered—and things aren't going to change anytime soon. There is certainly a strong reason to consider the medical profession a caring one, but even at the best of times doctor visits involve waiting, being resigned to only ten to fifteen minutes with the doctor, and getting treatment based on a snapshot of the situation.

There is an alternative. You can accept the role of the healing self. Consider the qualities of an ideal doctor, which would include the following:

Patience

Sympathy

An open mind

Vigilance over changes in the patient's condition

Close monitoring

Detailed knowledge of the patient's history

Thorough medical knowledge and expertise

Only the last item on the list is exclusive to the medical profession. Everything else is something you can provide yourself, either through self-care or in conjunction with a good doctor. Certain things like constant monitoring are available only to you (or by being admitted to the hospital). Most of what's on the list are things you are probably doing already, even though you aren't aware of acting as a healer. Maximizing them will be very important, because awareness needs to be an everyday habit, even a skill.

By the same token, the bad qualities we'd hate to see in a doctor are often present in how we treat ourselves from day to day. Millions of people approach their health with one or more of the following:

Indifference

Denial that pain and other symptoms need attention

Worry and anxiety

Lack of information

Guesswork

Undertaking needless or ineffective treatments

Clearly these are things that everyone wants to avoid, but we fall into self-defeating responses all the time. We worry needlessly or pretend that nothing hurts. We guess at what's wrong and then reach impulsively for something we hope will work—usually this means grabbing a bottle from the medicine cabinet or kitchen cupboard. Most of the time the impulse is temporary, so we go back to waiting and worrying.

You are in a position, starting now, to adopt the role of self-healer. By going deeper into the power of awareness, you can activate the hidden potential of the healing system that you already depend upon every day. We hope this all sounds exciting, because some major life changes lie ahead. But first we need to make very clear what this book is *not* about.

A Realistic Baseline

We won't show you how to overcome a chronic illness like arthritis, type 1 diabetes, or congestive heart failure.

We don't have cures for incurable diseases like Alzheimer's.

We aren't promising a cure for cancer.

Nothing we advise lies outside proven medical practice—we aren't talking about faith healing, placebos, or magical thinking.

Once you have developed symptoms or a full-blown disorder, you must seek qualified medical care.

Where Are You Now?

Some will be disappointed that this book isn't about curing a full-blown illness on your own. But the advantages of the healing self are immense, because you learn how to consciously remain in a state of well-being that increases for your entire life. As big as this concept is, healing comes down to personal experience today, tomorrow, and the day after. To that end, we are asking you to pause and take two quizzes. The first quiz will assess where you are now—in other words, your starting point for your healing journey. The second quiz will assess how great your potential is—how far healing can take you.

QUIZ #1: WHERE ARE YOU TODAY?

For each question, consider your experience in the past month. Mark each item according to how often an experience has occurred, as follows:

1 = Not at all or once at most

2 = Sometimes

3 = Fairly often

4 = Often

___ I was depressed.

___ I felt worried and anxious.

___ I had to go to the doctor's office.

___ I was in pain but didn't go to the doctor's office.

___ A chronic health problem was present.

___ I ate the wrong foods, fast food, or junk food.

___ I was under pressure.

___ I felt stressed out.

___ I had trouble getting to sleep.

___ I didn't get enough sleep.

___ I didn't control my weight.

___ I had a headache.

__ I had a backache.

__ My relationships didn't go well.

__ I got seriously angry.

__ I neglected exercise and physical activity.

__ I had self-doubt or self-esteem issues.

__ I felt lonely.

__ I felt unloved or uncared for.

__ There were troubling family issues.

__ I was worried about the future.

Assessing your answers

This quiz doesn't lead to a total score—each individual answer is what we will focus on. If you have lots of 3 or 4 responses, your life over the last month has been a struggle. Most people, however, will have some 3 or 4 answers no matter how well their lives are going.

Hold on to your results, and take the quiz again after you finish reading this book. Take it every few days or weekly after you adopt a healing lifestyle. If your answers improve, you have proof and motivation that such a lifestyle really works.

QUIZ #2: YOUR HIGHEST POSITIVE EXPERIENCES

Healing is a holistic process, and the healing self opens the way for higher experiences that make life more joyous and meaningful. We want to know how many higher experiences you are having now. For each question, consider your experience in the past month. Mark each item according to how often an experience has occurred, as follows:

1 = Not at all or once at most

2 = Sometimes

3 = Fairly often

4 = Often

___ I felt contented inside.

___ I expressed love openly to someone else.

___ I felt free and liberated.

___ I saw myself without blame or judgment.

___ I was appreciated and praised by someone at work or in my family.

___ I felt inner peace and tranquility.

___ I felt myself to be part of a larger plan or vision.

___ I experienced a loving gesture toward me.

___ I had a spiritual experience.

___ I felt lovingkindness and compassion.

___ I forgave someone.

___ I forgave myself.

___ I let go of something negative from the past.

___ I developed an emotional bond with someone.

___ I felt blessed.

___ I sensed what I'd call a divine or holy presence.

___ I was lighthearted.

___ My faith in human goodness was affirmed.

___ I felt blissful or ecstatic.

___ I saw or experienced inner light.

___ I experienced pure Being or unbounded awareness.

___ I saw the sacredness of another person.

___ I meditated, prayed, or did another contemplative practice.

___ I felt creatively inspired.

Assessing your answers

As before, this quiz doesn't lead to a total score—each individual answer is what we will focus on. If you have lots of 1 or 2 responses, your life over the last month has probably felt mundane and uninspired. Most people, however, will have some 1 or 2 answers no matter how fulfilled their lives are—access to higher experiences is still waiting.

As with the first quiz, hold on to your results, and take the quiz again after you finish reading this book. Every few days or weekly after you adopt our suggestions for a healing lifestyle, look at your answers again. If they improve, you have proof and added motivation that higher experiences aren't rare or random. They are accessible through your healing self anytime you want.

Now you have a better idea of what getting real about your health actually entails. You've discovered the concepts that are critical for moving away from the gambling model for staying healthy. Realizing that consciousness is the key now places you on the threshold of transformation. There are many details to explain, and the following chapters describe the nitty-gritty of our new model. Yet nothing is as important as knowing that the healing self is real. It's as close to you as the next breath and as vital as the next heartbeat.

2

Who Stays Well and Who Doesn't?

The beauty of a whole-system approach is its naturalness. All the most basic things we do to stay alive affect the whole system. We breathe, we eat, we go to sleep. Advanced medical science delves into these processes extensively, and the deeper the research goes, the more complex eating, breathing, and sleep turn out to be. But this shouldn't cloud a simple fact: the people who manage to stay well their entire lives, and who enjoy the highest state of wellness, are people who have no trouble getting eight hours of good sleep a night, who eat a balanced nourishing diet that maintains a healthy weight, and who breathe easily—that is, they are not burdened by stress and anxiety.

Millions of us cannot say that we've mastered the most basic whole-system processes. Somehow the naturalness of staying well has escaped us. How can this be? By analogy, think about a self-driving car. Long dreamed of by engineers, such a car has now become feasible, and its advent has been greeted with both optimism and panic. To the optimists, the self-driving car will be a boon to safety. Equipped with artificial intelligence and sensors that maintain a constant state of vigilance in 360 degrees, a driverless car would be able to detect potential dangers on the road almost instantly, far quicker than even the best human driver. But what if these safety mechanisms fail? That's where the panic comes in.

Being driven into an accident by a machine you aren't controlling feels like a nightmare.

In practice, therefore, a self-driving car needs to include a means for the human driver to intervene and take control. Judgment calls always occur in traffic situations. Not many of us, at this stage at least, would be willing to relinquish all control to a machine. Possibly we never will, given the risks to life and limb.

Much the same predicament relates to our bodies. Although perfectly self-regulating in its machinery—a simplistic term but useful in this case—the body is under dual control. At the outset of this book we mentioned the example of breathing. Whether you pay attention or not, you breathe in and out automatically; it's a basic survival mechanism. But anytime you want, you can intervene and breathe a different way, faster or slower, deeper or shallower. Because the body operates as a whole system, your interventions aren't local—a different style of breathing could be linked to a panic attack at one extreme and a mindful yoga practice at the other. Which means that every intervention has the possibility of moving you away from your natural state of wellness.

Apparently, millions of people have done just that. The signs are obvious in dozens of ways—poor sleep, chronic lifestyle diseases, obesity, anxiety, and depression head the list. The healing response is compromised by a huge disruption like pneumonia or polio, but these devastating events are becoming rarer and more curable. The real threat to healing comes from the daily interventions we make that have negative or unforeseen consequences; these are the raindrops that can eventually cause a flood.

The healing response holds no judgments; it adapts to every choice you make, positive or negative. Your cells are chemical factories that alter their production line depending on the input you give them, which acts like a directive from upper management. Because everyone's life is a mixture of good and bad choices, everything in life must be viewed as either raising your state of wellness or lowering it. Our cells, all the way down to the genetic level, tolerate our indulgences but are paying the price for them as well.

The solution is to use the body's dual control as a tool of healing. In the most basic terms, there are two kinds of healing going on in every person right now.

Automatic healing, which everyone inherited in their genes through millions of years of evolution.

Conscious healing, which covers every opportunity to assist and improve automatic healing.

Any experience is a possible candidate for healing. The simple fact is that a day without any sensation of physical pain doesn't escape the following experiences:

Feeling depressed, helpless, or hopeless

Worrying about the future

Feeling anxious, fearful, or unsafe

Being stuck in old behaviors and bad habits

Low self-esteem

Lack of fulfillment

Troubled relationships

Feeling lonely, shut out, and unappreciated

Leading a life without much purpose or meaning

Guilt and shame from old traumas and wounding

Who can say that something on this list isn't afflicting them right now or didn't occur in the past? According to a recent survey, one in six American adults is reported to be taking psychiatric drugs. Relieving symptoms, as we saw, doesn't get at the real cause of a disorder like depression. Research has attacked the cause of depression through brain scans, to see if a specific area of the brain is involved; through genetic profiling, to discover if there is a unique "depression gene" or group of genes; and through psychiatric evaluation, hoping that a pattern of behavior leads to the onset of depression.

Yet no specific cause has been found along any of these paths. The most accepted conclusion, in fact, is that each person's depression is unique, exhibiting an array of psychological, physiological, and genetic traits. Depression is linked to personal experience and how you react to it. Reading about bad news in the paper doesn't automatically trigger the same reaction in different people, who can respond with a huge range of reactions from indifference to deep depression. The same variability applies to anxiety, which is why one person collects spiders as an enjoyable hobby and another person is deathly afraid of them. Are you in a troubled relationship? Does your life seem to lack purpose and meaning? It's not the fault of medical treatment that such a huge range of perceptions is untreatable. These causes of suffering have no drug cure; they don't even fit the standard medical model of disease, which stubbornly insists on physical changes as the "real" cause of illness.

As a counter to this bias, impressive studies have shown that invisible subjective states can have a powerful effect on the body. For example, researchers at the University of Texas Medical School looked at mortality rates among a group of men and women who had received open heart surgery, including heart bypass and replacement of the aortic valve. If you take the routine medical approach, the reason someone dies six months after open heart surgery while someone else doesn't must come down to a physical difference. But the team headed by Dr. Thomas Oxman took an unorthodox approach. They asked these patients two questions about their social situation: Do you participate regularly in organized social groups? Do you draw strength and comfort from your religion or spiritual faith?

These are simple yes-or-no questions, and when assessing the answers, the researchers excluded the typical risk factors for dying after heart surgery, including age, severity of the disease, and severity of a previous heart attack. With these factors zeroed out, the findings were startling:

> A person who answered *Yes* to both questions had less than a 5 percent chance of being dead six months after their surgery.

> A person who answered *No* to both questions had between a 20 percent and 25 percent chance of being dead six months after the surgery.

Overall, being socially supported and taking comfort from your faith makes you seven times more likely to survive major heart surgery than someone who has neither of those things in their life. This outcome is almost certainly the only sevenfold difference in any risk for heart mortality, even bad cholesterol levels, high blood pressure, and a genetic history of heart attacks in the family. While asking somebody if they belong to social groups like a club or church is an objective measure, the question about religious or spiritual faith is entirely about how the person *feels*.

How you feel is entirely subjective, but just as important, it's an activity in consciousness, a small indicator of your self-awareness. The support for conscious healing couldn't be more obvious.

Marge's Story: Awareness Comes First

Much of the healing response remains a mystery. No one really knows—or can predict in all cases—why one person gets sick and another doesn't. The hidden reasons exist in a shadow zone beyond the physical.

Some people provide living proof that consciousness-based healing, as a way of life, works. Consider an elderly woman named Marge, who proudly remained active and self-sufficient until she was ninety-one. Marge still lived alone in her apartment, cooked for herself, continued to drive her car, and hired a house cleaner only for heavier chores. Marge's health was unusually good, in that compared with an average of seven prescription drugs taken by seniors over age seventy, she was on just one medication, for high blood pressure.

In an aging population, more people will be eager to know Marge's secret. Is it good genes for longevity? To date, no definitive research has clearly found such a gene or group of genes (although there are strong clues, as you'll see a bit later). Typically, if your parents both survived to eighty, your lifespan is expected to be three years longer than average, not a great advantage.

As a statistic, Marge did have certain advantages. She was born into a well-to-do family in Cincinnati, which meant that she had good medical care, not that this would have saved anyone from serious childhood illness in 1920, when she was born—antibiotics, for example, still lay in the future. She was lucky not to contract tuberculosis, polio, or scarlet fever. An absence of serious childhood diseases is associated with longer life.

But in her mind, none of these factors was the deciding one.

"I had a difficult marriage to an artist in New York," she recalls. "We were both strong willed, and we fought a lot. Most of my energy was spent on him rather than on my three sons. I'm not proud of that, and as much as I loved my boys, I could be harsh with them."

Looking back, Marge realizes now that it took decades for just one psychological trait—her anger—to have drastic effects on her life.

"I divorced when my boys were barely teenagers. One went off to boarding school, and the other two chose to live with their father, who was so angry with me in general that he fought to keep all our money for himself and his sons. Suddenly I was completely alone and bewildered over how my life could be overturned so drastically."

Marge fought with depression. It became obvious that her sons had been scarred by her hot temper growing up. "I forgot what made me so mad the minute things calmed down, but they didn't. They actually became afraid of me, their mother."

Up to this point in her story, nothing would suggest that Marge would live longer than average—perhaps the reverse, if her depression had become chronic and affected her health status. Then a single factor changed her life—Marge became a Buddhist. To her, this decision created an inner transformation.

"I found a Zen master through a friend," she recalls. "I can't even tell you why I gave Buddhism a try, but once I started to meditate, two things happened. First, by becoming calmer, my bad temper wasn't triggered by every little thing. Second, I saw something real about myself. Deep down, I was very anxious about being alone. All the drama I kicked up was a tactic to get people to pay attention to me, and that defended me from how alone I really felt."

Today, at ninety-six, Marge lives in a retirement home at the lowest level of their assisted-living program, which means that somebody looks in on her several times a day and helps her bathe. Her medications haven't increased. She goes down to lunch and dinner on her own and takes excursions with friends to eat out once a week. Two things only are areas of difficulty.

"My hip replacement, which I got in my seventies, has begun to wear out, so I've decided to use a wheelchair rather than walk a long distance. And my sons are still wary of me. Having an angry mother isn't something they have been able to overcome easily. That's the only sorrow that lingers in my mind. Otherwise, I'm at peace."

Marge was fortunate to begin meditating so long ago, because the late 1970s was really the first time mainstream medicine began to produce studies showing that meditation was associated with positive health outcomes such as lower blood pressure. "Relaxation" was the recognized tag word for other benefits, such as reduced stress and anxiety. Skip ahead forty years, and meditation became more popular and accepted. Today, the whole-system approach dissolves all the artificial barriers between mind and body. The realization that every experience has a mental result and a physical result is gathering greater force every day.

Take a simple fact, that grief lowers a person's immune response. Grief is a drastic mental event, a source of psychological pain. When someone is still seriously grief stricken six months after a death in the family, which occurs in about 10 percent of bereavements, their condition is known as traumatic grief. Studies on those who suffer from traumatic grief indicate "global impairment" is very likely, meaning in layman's terms that almost anything can go wrong with their health.

A study of 150 widows and widowers, for example, found that "the presence of traumatic grief symptoms approximately 6 months after the death of the spouse predicted such negative health outcomes as cancer, heart trouble, high blood pressure, suicidal thoughts, and changes in eating habits at 13- or 25-month follow-up." (Rudy remembers that after his father died at the age of forty-five, when Rudy was seventeen, it took years for his mother to overcome her grief and live a normal life again.) If you absorb what this all means, certain people, for unknown reasons, are hit harder by grief than others. The natural recovery time for grief doesn't heal them, and up to two years later, they are at risk for all kinds of disorders, both mental and physical. Other studies have found similar results relating to disturbed sleep, low self-esteem, and sad mood.

Traumatic grief throws into high gear the power of the mind-body connection. Although medical science can tell us many things on the physical side about cancer and heart disease—and going deeper can even spot the chemical imbalances that might show up in someone who suffers from traumatic grief—there is no *cause* that triggers this kind of long-lasting grief, no knowledge about *why* the healing system failed, and little understanding about the *purpose* and *meaning* of grief in the first place. (Other mammals do not appear to grieve, with suspected exceptions like elephants and domesticated dogs. If a deer is shot by a hunter in a herd of deer, the rest of the herd is disturbed briefly before going back to normal grazing.)

The italicized words in the last paragraph—*cause, why, purpose, meaning*—point to an undeniable fact. Humans live for a purpose, and when the purpose is damaged—your beloved spouse dies—grief can make life seem meaningless. Every cell in the body gets the message in chemical form. The chemicals are the physical evidence of grief, but the loss of meaning isn't chemical—it's human in the larger sense. As painful as grief is, if someone felt no grief at the loss of a spouse, that would be considered strange—some might say behind their backs that they are *heartless*, another very human word.

Invisible Healing

The healing self is the part of us that deals with invisible causes, the why of who gets sick and who doesn't, and the purpose and meaning of being alive. Healing isn't mystical just because it's invisible. Someone who has never thought about the healing response will in all likelihood want to be happy, and one key to overall happiness is feeling loved. Is it really possible that your cells can feel loved, too? Before reacting to what seems like a ridiculous assertion, consider the following study.

Yale researchers looked at 119 men and 40 women who received the most accurate test for detecting blockages in the coronary arteries, known as coronary angiography. (It's an anxiety-provoking procedure for many people, although relatively noninvasive. Typically a narrow catheter inserted in the forearm is threaded into the arteries of the heart. A dye is injected that will show up the interior of the artery using a CT or MRI scan. In this way the size of the vessel's opening or blockage can be seen directly.) Patients who told the researchers that they felt loved and emotionally supported generally exhibited less blockage in their coronary arteries, the main cause of heart attacks and strokes.

There are other risk factors that predict the presence of heart disease, such as diet, exercise, smoking, and family history, but even when these were taken out of the equation, the feeling of being loved and emotionally supported was a predictor of who would have more or less arterial blockage. A study of 131 women in Sweden came to the same conclusion.

But perhaps the most striking research was based on asking a single question. A team at Case Western Reserve University surveyed 10,000 married men with no history of angina pectoris, the typical chest pain associated with heart disease (although heart attacks can occur without this previous symptom).

As expected, the men who scored highest on the familiar risk factors for heart disease, such as high cholesterol, hypertension, and older age, were more than twenty times more likely to develop angina over the next five years. Then the researchers asked a simple question: "Does your wife show you her love?" The men who answered *Yes* were less likely to develop angina even when they had high scores on known risk factors. The reverse was also true. A man with high risk factors who said his wife didn't show him her love was almost twice as likely to develop angina.

As with traumatic grief, taking the mind-body connection seriously is enough to explode two of the most common assumptions driving medical care:

1. Healing is physical and happens automatically.
2. When the automatic healing response breaks down, the only thing a doctor can do is to intervene with drugs or surgery.

From opposite extremes of the emotional spectrum, love and grief cross the boundary between mental and physical. Heart disease is treatable with drugs and surgery, but these can prove ineffective in someone who feels isolated, lonely, or unloved. The unpredictable physical effects of traumatic grief aren't treatable by drugs and surgery—after all, you can't give a pill for everything that might go wrong thirteen to twenty-five months from the onset of grief. By ignoring the healing self, a physician in everyday practice has left out a key part of health and healing.

Basic Awareness

Taking everything we've said so far, the benefits of conscious healing are there for the asking. But for many people *conscious* simply means that you aren't asleep or knocked out. They have the same potential to be aware as the most advanced yogi or monk, but no one taught them how to use this ability. Take three people and sit them in the same room, then ask them what they are aware of. You'll receive random answers that won't necessarily overlap. One person is aware of a smell in the room, another of the wallpaper, the height of the ceiling, and so on, depending on what is being noticed at the moment. Less likely is that one of them will be aware of an inner state—thoughts, moods, sensations. Only if you blatantly change the environment, like turning up the temperature to eighty-five degrees, will everyone be likely to mention the same thing.

Spiritual practices in yoga and other Eastern traditions are actually about honing random awareness to make it sharper, turning an innate ability into a skill. Before they are aware of anything "out there" or "in here," those who have trained their awareness will universally say that they are self-aware. The average person is also self-aware. You can't have a sense of "I" without it. But self-awareness is only one piece of the swirling, random, unpredictable activity going on in the mind.

Awareness skills don't have to be associated with spirituality or the East. They can be used to improve your quality of life. That's where the healing self becomes practical in any situation, at any moment of the day, and with any religious background. It monitors the signals that indicate your immediate state of well-being, here and now. These levels include the following:

Knowing how you feel physically. This involves being open and sensitive to the signals your body is sending you.

Knowing how to interpret these signals. This involves acceptance of your body as your greatest ally, not a source of distress.

Knowing what is happening inside you emotionally. This involves giving up on denial, wishful thinking, fear, and repressing your emotions.

When someone casually asks, "How are you?" we usually give an equally casual answer, but the healing self takes the question seriously. By knowing what is actually going on, you are beginning the process of self-healing. A wearable device can buzz you when your heart rate jumps, your blood pressure rises, or your breathing becomes erratic; these are useful indicators, certainly. But only you can respond to the signals and begin the healing.

As a practical example of basic awareness, here's what you can do almost effortlessly at work.

Mindful at Work: Seven Self-Aware Things You Can Do Right Now

Adopt any or all of the following tips to counter the invisible negative influences that afflict the typical workplace.

1. In Eastern traditions, awareness should be one-pointed, which means that you keep your attention in a state of relaxed focus. Don't multi-task, which divides your attention and has been proved to reduce efficiency at work.

2. To keep your focus relaxed instead of tense, do what you can to work in an area that's quiet and relatively free of interruptions. So that your coworkers won't feel that you are unavailable, take time twice an hour to circulate, be in contact, and let it be known that you want personal interactions. In this way, your alone time is likely to be more respected.

3. Awareness should be in the now. To stay in the present moment, don't let small demands pile up. Immediately take care of anything that takes five minutes or less. If you make this a habit, your time management will improve, sometimes dramatically, and you won't get to the end of the day complaining that you didn't have enough time to do everything you needed to do.

4. Be mindful of your body and its needs. At a minimum get up out of your chair, stretch, and move around at least once an hour.

5. Be mindful of your core or center. When you feel frazzled, find a quiet place where you can close your eyes, take some deep breaths, and become centered again. Some people find that centering works better if they put their attention in the region of the heart.

6. Remember to breathe, because breath connects many body functions, including heart rate, blood pressure, and the stress response. At least once an hour, do a few 10-count breaths, as follows: breathe

in to a count of 4, hold for a second as you relax into the feeling of the in breath, then breathe out to a count of 6. (Make sure your pace feels comfortable, not so slow that you gasp after a few breaths.) Typically, your breathing rate will slow down from 14 breaths per minute to 8, with an accompanying feeling of calmer mind.

7. Be mindful of your ultimate purpose, which isn't to meet a deadline but to create a day with happiness in it. Psychologists have found that people who lead the happiest lives follow a strategy of having happy days. Whatever makes you genuinely smile counts as a happy experience.

These same practices are also effective outside the workplace. Yet we typically spend over one-third of every weekday at work, and sometimes far more (the average white-collar worker who brings work home is estimated to spend sixty hours or more per week). It can be a challenge to remain self-aware under workplace pressures and demands. But the benefits are considerable—if you can stay centered and focused without being frazzled by the mental "noisiness" that crowds into a typical workday, you will be genuinely meditating in the midst of action, one of the primary goals in every wisdom tradition. Leaving spiritual issues aside, being self-aware is a major component of a healing lifestyle.

3

Nothing Is Better Than Love

The research showing that people who feel loved are more likely to enjoy healthier hearts than those who feel unloved is important. It puts real science behind something we all know personally: Love is the healthiest of all emotions. It sustains life at a level of trust, joy, and compassion unrivaled by anything else. A great poet from India, Rabindranath Tagore, declared that love was no mere emotion but a cosmic force. To be raised in a loveless household is the cruelest fate a child can endure, as the following story poignantly illustrates.

Patrick, now in his early thirties, didn't believe that he had suffered emotional damage as a child. He only knew that when his mother said she loved him, he couldn't expect her to hug him or even touch him. Her distance was a constant from an early age.

"I was hospitalized when I was five to get my tonsils taken out," Patrick recalls. "It was Valentine's Day, and I was in a ward with other kids. Their mothers came by with cards and candy, but my mother didn't. But that's the funny thing. What I remember is turning my face to the wall and holding a pillow over my ears so I couldn't hear the other kids and their moms. I carried a grudge around for years, and then a peculiar thing happened. One day my mother and I were having lunch, and my curiosity got the best of me. I asked her why she never came to see me in the hospital, and you know what she said?

"She told me she arrived a bit late when visiting hours started, and she found me curled up in the bed crying. She comforted me, she said, but that part I don't remember, only the feeling of being alone and forgotten."

Young children, as many psychologists attest, form powerful beliefs about their upbringing that don't always match the facts. It has taken medical science a long time to stop believing that only facts, as measured by medical tests and diagnoses, are the sole component of being healthy. Beliefs matter, even when those beliefs are totally subjective—we all believe the stories we tell ourselves. This story begins with the messages our parents send in childhood.

Patrick's parents sent a message that told him he was on his own. Being aloof and unaffectionate themselves, they treated this as normal. But children need to feel closely connected to a parent who is loving and protective. This is an evolutionary trait millions of years old. In a famous experiment baby monkeys were separated from their mothers and soon became restless, anxious, and insecure in their behavior. When given an artificial mother constructed out of wire mesh with padding around the torso, the babies immediately attached themselves to it, clinging for comfort.

In humans the effects of poor bonding are just as devastating, even though we have superior abilities to adapt even to the worst conditions. In Patrick's case, believing that he was on his own led to what is known psychologically as "loose attachment." In layman's terms he didn't feel safe, valued, and protected. In his mind, true or not, if he was in trouble no one would be there for him. That's an exaggerated, black-and-white notion—no doubt his parents would have been shocked to hear it—but kids' ideas tend to be like this, based on indelible emotional experience.

"I was lucky in a way," Patrick recalls with a faint smile. "I got very good at being independent. Everyone commented on how I was like a little adult even when I was seven or eight. I was proud of it. I became a superachiever because that's what adults do, and this went on for a long time."

When he started dating in his teens Patrick didn't fill the hole in his heart; he had no idea how to. Getting close to someone felt alien, in fact. Keeping true to his story that he had to look out for himself, he was motivated mainly by his growing sexual urges. Girls tended to go along with this, and if they started to want more and the relationship verged on getting serious, Patrick found excuses to start a fight or act so cold that the girl, bewildered and hurt, walked away.

By the time he got to college, studying computer science, Patrick's isolation, emotionally speaking, was baked in. His ability to take care of himself was unquestioned. What he didn't realize was this: He took care of himself because he believed that nobody else would. He had no model for a nurturing, protecting love.

The story could have stopped there. Fortunately for him, it didn't. In graduate school he met a girl, and she was different. Patrick's totally rational worldview collided with love at first sight.

"To be honest," he says, "I thought those words were just clichés. The first time I met Fran, I saw her for only a second outside the department while she was talking to some friends, but I was riveted. There was just something about her. I got up my nerve and introduced myself. She was friendly and smiled. Nothing else happened, but I went home, and Fran was all I could think about.

"She agreed to go out with me, and things just unfolded from there. Without warning I had become a fool in love. I died inside if she had a change of plans and couldn't see me. I kicked myself every day to make sure it was true—I'd fallen in love with the most beautiful woman in the world."

Despite the undoubted experience of falling in love at first sight, and abundant evidence that love creates powerful physiological changes, the whole phenomenon remains mysterious. Do chemical changes in neural activity mean that Patrick's brain fell in love, or did *he*? In a whole-system approach the two are inseparable. Where healing is concerned, there are some deep issues that cross the mind-body boundary:

How and why does love promote physical health?

When infatuation turns into lasting love, what can this do for our well-being?

If love becomes deep enough, does it open the door for higher consciousness?

The human experience testifies that love has a unique power in all these areas, and if we investigate deeply enough, there are answers for why this is so.

Love Reaches Deep and Far

We live in an age in which such an overwhelming experience, one that transforms someone's entire being, is explained as biochemistry. But even with sophisticated brain scans and measurements of hormone levels and the like, what's missing is the *meaning* of falling in love. That meaning is all embracing. In the terms we're using, love is a whole-system event. The studies that indicate how coronary arteries respond to love, or the lack of it, are only the tip of the iceberg. We carry a deep evolutionary imprint at a genetic level. To quote psychologist Barbara Fredrickson, "Somewhere in our brains we carry a map of our relationships. It is our mother's lap, our best friend's holding hand, our lover's embrace—all these we carry within ourselves when we are alone. Just knowing that these are there to hold us if we fall gives us a sense of peace."

What's most significant is that even when someone is alone, sitting quietly and passively, they aren't really alone. Inside they carry a map constructed from all the relationships experienced since infancy. This, too, is a whole-system phenomenon. Each moment in a relationship is a tiny piece being fitted into the whole map as it changes and shifts.

To see how this works, *give the first answer* that comes to your mind when you read the following statements:

My mother loved me enough.	YES	NO
I'm glad I had the father I had.	YES	NO
I trust where I am right now.	YES	NO
My current relationship is in a very good place.	YES	NO
I have a good friend I'm closely bonded with.	YES	NO
I like being emotional.	YES	NO
I tend to show how I feel.	YES	NO
Others feel safe confiding in me.	YES	NO
I am a nurturer.	YES	NO
I feel as if I belong.	YES	NO

There are no right or wrong responses to these statements. But if you answer quickly without pausing to give what you believe is the "right" answer, your responses come directly from your inner map of love and relationship. You may be happy or shocked by your responses, and we will be showing you how to improve your inner map in many ways. Here just be aware of how an inner image you probably never think about is affecting you, not incidentally but as the story of who you are, involving the whole person.

A whole-system approach would predict that love, or the lack of it, would have multiple effects, and it does. In terms of biochemistry, a number of significant changes take place when someone falls in love. Levels of brain neurochemicals such as dopamine and serotonin go up, together with hormones like cortisol and follicle stimulating hormone. These are the earliest changes of falling in love. Ironically, the last two are indicative of stress caused by the arousal between the sexes. In other words, there's a chemical basis for why romantic love brings joy and pain. Shakespeare intuited a fact of neuroscience in *A Midsummer Night's Dream* with the famous line "The course of true love never did run smooth." Even more intriguing is that testosterone levels go down in men and up in women as we age, leading to changes in temperament that render two members of the opposite sex just a bit more alike.

Even more far-reaching are the possible effects of love on the healing response. How well your immune system works is a crucial factor, and it is well documented that emotions change the immune system. Drs. Janice Kiecolt-Glaser and Ronald Glaser looked at couples who had been married for a long time, on average forty-two years, and found that those who constantly argued had a decreased immune response. If this finding seems to be a sad comment on older marriages, the same effect occurs very quickly, too. A study of couples who were on their honeymoon showed that newlyweds who showed hostile and negative behavior when asked to discuss the topic of marital troubles had a decreased immune response.

By now, having read this far, you won't be surprised that the body responds to positive and negative emotions. But the speed of this response may still surprise you. A pioneering study by psychologist David McClelland and his team at Harvard asked students to watch a film on the work being done in the slums of Calcutta by Mother Teresa, the Catholic nun who gained worldwide fame by caring for the poorest abandoned children. (As a control, a second group of students watched a neutral documentary on another subject.) On average the students who watched the Mother Teresa film showed increased levels of antibodies on the spot, along with decreased measures of stress like reduced blood pressure.

This finding is impressive for showing how the body responds here and now to emotional experiences, but McClelland wondered why some of these students actually had a decreased immune response to watching Mother Teresa's good works. As a follow-up, all of the original study group were shown a photo of a couple sitting on a bench by a river. When asked to write a story about the couple, some students described them as loving, supportive, and respectful, while others wrote very different stories in which the couple was unhappy, manipulative, and deceitful. The students who showed the biggest decrease in immune response in the first study happened to be the same ones who wrote the negative stories in the follow-up. The implication is powerful: ingrained ideas we carry around inside us wind up defining what relationships are all about, even when those ideas aren't aligned with the truth. They forcefully impose their own interpretation instead.

Going back to Patrick, he and Fran had entered the first phase of romantic love, which is infatuation. In that state, everything about love is so powerful that reality changes. Your beloved is the most beautiful person in the world. In the presence of the beloved you have entered paradise. Under love's spell, the whole world looks brighter and is inhabited by wonderful people. To a strict rationalist, these are illusions. And in fact, infatuation is temporary; its intoxication wanes, giving way, if the person is lucky, to more stable stages of love. In these later stages other neurochemicals, such as endorphins (natural opiates), oxytocin, and vasopressin follow predictable patterns in response to the back-and-forth of lovers. But is falling in love just a set of chemical experiences?

There's an important factor rationalists don't consider. Falling in love gives a *more realistic* view of life by placing us in contact with our true self. We fall accidentally and temporarily into a state of expanded awareness that is exalted by the great mystic poets, who connect intense human love with divine love. The beloved Persian poet Rumi exults:

Oh God, I have discovered love!
How marvelous, how good, how beautiful it is! . . .
I offer my salutation
To the spirit of passion that aroused and excited this whole universe
And all it contains.

There is no doubt that love can expand into this higher dimension, where the whole person is healed in the most profound way. The connection between body and mind is undeniable, but that's also true in an everyday instance of falling in love. The following experiences pertain to love wherever it is felt:

Feeling renewed

Bonding at the level of the heart

Feeling protected and safe

Emotions of joy, exhilaration, and upliftment

A more open heart, extending empathy and sympathy to others

Feeling physically lighter

Sensing energy or light coursing through the body

There's no distinction here between what a saint would describe and a Patrick who discovers love for the first time. He and Fran didn't last as a couple beyond a year. Like everyone who has passed through the stage of infatuation, they had ego needs that weren't the same. Settling down to love while negotiating the demands of "I, me, and mine" poses its own challenges. But Patrick learned the most valuable lesson of his life, that he was lovable and, along with this, that he could love.

Humans are not biological robots. We live for meaning, for the personal value of every experience. The body metabolizes our experiences and sends the message to every cell, while the mind, in its own domain, processes experience in terms of sensations, images, thoughts, and feelings. Nothing fuses the whole-system effects of love and non-love like the human heart, which needs to be understood as more than a physical organ.

A Whole-System Vision of the Heart

The heart offers one of the best examples of how the whole-system approach makes the most sense. Heart disease is the leading cause of death in this country for both men and women, which makes it a primary target for a healing lifestyle. Your heart is very responsive to how you feel emotionally and physically. There's almost no choice you make that your heart doesn't know about.

Yet very few people are really aware of this. Until they feel an alarming symptom like chest pain, they might think about heart health mainly in terms of cardio at the gym. Other disorders, such as breast cancer, attract more publicity and generate more fear among women, but statistically, this perception doesn't conform to the actual situation. Out of total deaths among American women each year, breast cancer accounts for 1 in 31, while deaths from heart disease are 1 in 3. Depression and anxiety are associated with increased risk for exercise-induced heart attacks. In contrast, greater levels of positive emotions are associated with decreased risk. Being heart healthy is important. But in our whole-system approach, the physical organ is only part of the story. The other part has to do with one's attitudes and outlook.

Even if someone holds that a purely physical approach is adequate, that side of the equation isn't completely well understood. For example, a study in the early 1950s looked at the hearts of young soldiers who were casualties in the Korean conflict. America was just becoming aware of an epidemic of heart attacks afflicting men between the ages of forty and sixty. No one knew what was causing this alarming spike in premature heart attacks. The fashion for mainly blaming cholesterol hadn't yet begun, and there were no drugs like cholesterol-lowering statins to prevent heart disease.

Into this baffling environment, the hearts of young soldiers told a grim story. A high percentage of them displayed considerable plaque blocking their coronary arteries. Plaque is hardened fat mixed with minerals and caked blood that can shut off the heart's own oxygen supply. When an artery is strangulated, the heart muscle goes into convulsions, entering into a full-blown heart attack. It was assumed that plaque buildup took decades, gradually increasing the risk of heart failure.

However, these were men in their early twenties, and their arteries were sometimes blocked as badly as older men who suffered from heart disease. How did this happen? Just as mysteriously, why did the heart wait until a man was forty before a heart attack occurred? These questions remain unanswered today. The relationship between arterial plaque and all the possible factors that might cause it—diet, blood fats, stress, genetics, and microscopic changes in the walls of the coronary blood vessel—is extremely complicated.

The most obvious fact is that neither the soldiers nor the doctors who gave them a physical checkup had a clue that something serious was going on. (The advent of sophisticated tests like angiography lay decades in the future.) The typical chest pain associated with heart disease, known as angina pectoris, generally shows up late in the game, and it's also possible to have blocked arteries with no pain—in such cases, a heart attack occurs out of the blue. A healing lifestyle needs to be followed whether pain exists or not.

Even with so many open questions, once heart disease has been diagnosed, after a brief visit with a cardiologist, a battery of tests, the typical first step is a prescription drug to combat high cholesterol or elevated blood pressure. Lip service is paid to lifestyle changes, sometimes not even that. A patient's motivation to pursue diet and exercise generally isn't strong in the first place. A lifetime of settled habits is hard to turn around. If the patient's condition continues to worsen, some kind of surgical procedure is in the offing. The two most popular interventions are angioplasty and coronary artery bypass graft. Here's a thumbnail sketch of the "simplest" intervention, angioplasty, which in the United States is performed more than 600,000 times annually.

ANGIOPLASTY: MORE DOWNSIDE THAN UPSIDE?

WHAT IS IT? Angioplasty involves inserting a tiny balloon into a blocked artery of the heart to expand the artery. The theory is that by opening up the artery, better blood flow will be achieved to the heart, decreasing the risk of a heart attack. Routinely a wire-mesh stent (a short, narrow tube) is inserted to keep the artery open after surgery. The relatively low risk of angioplasty helped to fuel a dramatic increase from 133,000 procedures in 1986 to more than 1 million annually by the 2000s, leading to a $100 billion industry today when you include heart bypass surgery. As with all surgical interventions, however, angioplasty has pros and cons.

PROS: The intervention is not severe physically, involving a tiny wire over which a balloon is slid into the artery. Many times, one has no choice but to have this surgery for survival—after a heart attack, for example.

Angioplasty is quick and not terribly uncomfortable.

After a night of observation in the hospital, recovery is rapid. Patients generally resume their normal lifestyle.

The main purpose of angioplasty, to make heart patients feel better by relieving chest pain or offering psychological relief from anxiety, is often achieved.

CONS: Angioplasty doesn't cure the underlying disease, which continues to progress. The procedure must often be repeated, and stents replaced.

Significant extension of life expectancy is not generally the case, especially in older patients. (A strong exception in extreme circumstances is with patients who have recently suffered a heart attack.) The first clinical trials of angioplasty, in the early 1990s, revealed no survival benefit of elective angioplasty as compared with medication.

The serious immediate risk is that arterial plaque may be dislodged by the balloon, leading potentially to a heart attack (or a stroke if the stent has been placed in a blocked carotid artery in the neck), occurring in 1 to 2 percent of procedures. Arteries can rupture if the balloon is inflated too much. Various possibilities of infection are present.

Angioplasty is expensive, with costs varying widely. The cost is often not worth the results. A 2008 presentation at the annual conference of the American Heart Association (AHA) concluded that angioplasty relieves chest pain in some heart patients but "at a cost generally considered to be prohibitive as a routine initial management strategy." Despite this conclusion, more than 1 million Americans receive heart stents every year.

A far more serious intervention, heart bypass surgery, involves putting the patient on an external pump to maintain circulation while the heart is being operated on. This increases all the risks just covered with angioplasty and is even more expensive. We won't go into detail except to offer some highlights:

- Heart bypass surgery is more painful, takes longer for complete recovery, and still doesn't significantly increase life expectancy except in specialized cases where the primary coronary artery is severely blocked.

- Even then, because few patients heed the advice to improve their lifestyle, plaque can begin to damage the grafted blood vessel in a matter of a few months. (The first patient to successfully receive a bypass graft in 1960 was relieved of his angina symptoms for just one year.)

- The inventors of the procedure predicted, wrongly, that bypass surgery would be a rare intervention, useful in patients with imminent risk of heart failure. At present, however, more than half a million coronary artery bypass surgeries are performed in this country every year.

Of the many cons involving angioplasty and heart bypass surgery, the one that stands out is the first: the underlying disease isn't cured. In the 1980s, pioneering studies by Dr. Dean Ornish at Harvard Medical School conclusively showed that positive lifestyle changes can do more than prevent heart disease; they can heal it. Ornish's program of diet, exercise, meditation, and stress reduction, then considered revolutionary, actually opened up clogged coronary arteries, the first success in any form in reversing heart disease.

A lifestyle-based approach remains the only proven way to reverse the plaque that lines the coronary arteries in people at high risk for heart attacks. The inclusion of meditation in the program was considered daring and controversial at the time—the medical profession still harbored the prejudice that meditation was an esoteric Eastern religious practice and therefore had nothing to do with "real" medicine. Now it's become accepted therapy to recommend meditation for high blood pressure, anxiety, insomnia, and other disorders. But the original Ornish program was absolutely strict, demanding adherence to stringent dietary rules. For example, one rule limited all fat intake to, at most, a couple of tablespoons a day.

For the vast majority of people who have not been diagnosed with heart disease or suffered a heart attack, the search for an ideal lifestyle, one that heals and not simply prevents, remains open ended. Dr. Ornish has published many books and articles that explore healing in body-mind terms. Medicine will continue to compartmentalize mind and body, but as individuals seeking healing, we can't afford to. The original lifestyle research broke down the wall separating mind and body. Without them, the whole-system revolution couldn't have taken place.

The emotional states associated with the heart include some that every life would benefit from:

Empathy, which makes us feel what someone else is feeling

Compassion, which motivates us to extend lovingkindness

Forgiveness, which wipes the slate clean of old grievances and wounding

Sacrifice, which allows us to put someone else's good above our own

Devotion, which inspires reverence for higher values

None of these states is a term in cardiology, yet they have medical consequences. In the next chapter, we'll discover how recent breakthroughs are transforming matters of the heart. But here we want to reinforce the healing value of love. People thrive when they feel loved and languish when they don't. Love increases one's sense of self-esteem, which leads to taking better care of oneself. Love also alleviates stress, anxiety, and depression, which reduces chronic inflammation and risk for many age-related disorders, such as heart disease, diabetes, and Rudy's specialty, Alzheimer's disease. Love is a state of awareness, not a lifestyle choice. Ultimately what counts aren't the choices you make but the consciousness that keeps those choices going in a constant state of healing.

4

Lifeline to the Heart

The fact that emotions play a part in heart disease is only one piece in a complex picture. Where the common cold has a single cause, the rhinovirus, coronary heart disease (CHD) doesn't; it is surrounded by a cloud of risk factors, no one of them playing the leading role. Two people can develop heart disease or escape it while seemingly facing the same risks. This may surprise you, because in a word association game, the first word most people would come up with when thinking about heart attacks is *cholesterol*. The campaign to prevent heart disease expends billions of dollars on drugs that lower cholesterol levels in the blood. In the face of a nondrug program that actually reverses CHD, pioneered by Ornish and still viable, the proportion of people who turn to drugs instead is overwhelmingly high. Our aim is to inspire you to pursue a healing lifestyle, but people are habituated to depending on doctors and drugs almost without thinking. In the case of CHD, fixating on cholesterol has never been the whole solution because it doesn't fully address a complex disorder.

The cloud of risk factors surrounding CHD illustrates perfectly the huge advantage of a whole-system approach. If your heart is responding to how you lead your life, including your relationships and your emotional life, keeping your heart healthy should be all encompassing. Let's begin by sketching in how the cloud of risks came to be formed. As we mentioned in the last chapter, one of the great mysteries in medical history was the epidemic of premature heart attacks that struck America in the 1950s. Heart disease was traditionally considered fairly uncommon. At the turn of the century one of the leading surgeons in the country, William Osler at Johns Hopkins Medical School, declared that a physician in general practice would be likely to encounter a case of angina once a year. Jump ahead to the 1950s and doctors saw patients, predominantly men, complaining of chest pain on a weekly or daily basis. In 1900, pneumonia was the leading cause of death in the United States, a time when the average life expectancy was forty-seven. By 1930 heart disease had reached number one, where it has stayed ever since, and life expectancy was sixty.

What happened in between? The usual explanation is that people were living longer and heart disease rises sharply with age. With longer lifespans, a disease that was always prevalent was being unmasked. Better sanitation had played a major part in allowing people to live longer, and the germ theory of infectious diseases led to better prevention. Even when infectious diseases were massively reduced by the introduction of antibiotics, especially penicillin, no one anticipated that deaths from heart attacks among men ages forty to sixty, the range considered premature, would climb to alarming heights after World War II, resulting in an epidemic that reached its peak in the mid-1960s. Since then, deaths from heart attacks and strokes have steadily declined even though our life expectancy has kept growing.

A Cloud of Risks

The steady decline in deaths from heart attacks was not simply the result of controlling cholesterol. The most important factors can be quickly summarized:

- Many heart attacks were caused by infection of the heart (acute endocarditis), which could be detected with blood tests or echocardiograms and treated with antibiotics. Some researchers argue that this factor played the leading role in decreased heart attack deaths.

- Better treatment in hospitals raised survival rates after someone suffered a heart attack.

- In a combination of the above, patients with detected heart infections could be treated in the hospital, a much better environment for survival in case they happened to have a heart attack due to the infection.

What one doesn't see is an improvement in factors that started to build the risk cloud. These are also easy to summarize:

- Paul Dudley White, a prominent Harvard cardiologist, was appointed to be President Eisenhower's physician after Eisenhower suffered a heart attack in 1955. White was of the opinion that a shift in the American diet was a chief cause of the heart attack epidemic. Before and during the Great Depression, low incomes kept most Americans eating a high proportion of vegetables and small amounts of meat. With postwar prosperity there was an unprecedented rise in high-fat diets rich with meat.

- White, who is credited with starting the push for heart attack prevention, also pointed to the health benefits of exercise, as American life was becoming more sedentary.

- A third factor White stressed was weight control.

- Later on, as stress became better understood, the concept of the type A personality entered the popular culture. Heart attacks were linked to type A traits of being tightly wound, demanding, driven, and perfectionist, as opposed to the type B personality, which was more relaxed, accepting, and generally undemanding.

- Toxicity entered the picture when the ill effects of tobacco were being focused on. Although lung cancer was the primary target, smoking was also found in time to attack the lining of blood vessels, including the coronary arteries.

- A gender gap in heart attacks between men and women was traced in large part to the role of estrogen, which protected women from heart disease until menopause.

- Hypertension (high blood pressure) was found to aggravate heart disease by putting pressure on the lining of coronary arteries, exacerbating the tiny cracks where fatty plaques first begin to be deposited.

As you can see, this cloud of risks does not stand solely on cholesterol, so on the surface it seems peculiar that one factor in our diet—and a totally necessary chemical for cell structure—should emerge as the only villain. Inveterate cynics like to point out the huge profits amassed by drug companies pushing drugs, or the "silver bullet" mindset of Americans on the lookout for a pill that instantly solves the problem, abetted by the willingness of doctors to prescribe a cholesterol-lowering drug when they know that CHD is a complex disorder with multiple routes to prevention.

Yet cynicism doesn't lead to solutions, which is what we are after here. Controlling risk factors remains important. Despite the problem of noncompliance, more Americans now lead a lifestyle that's good for their hearts: they exercise regularly; consume less fat as well as less sugar (the latter, it is now suspected, could pose a bigger risk than saturated fats); meditate and do yoga; and do not smoke.

We will address these standard prevention measures in a later chapter, including the thorny complications of cholesterol. But making cholesterol the villain, and cholesterol lowering the panacea, isn't valid. The Medical Clinics of North America, after a meta-analysis (i.e., one that examines multiple studies) of four large primary prevention trials, suggested in 1994 that a 24 percent reduction in nonfatal heart attacks and a 14 percent reduction in fatal heart attacks could be expected because of cholesterol-lowering therapy.

So stubborn is our addiction to silver-bullet drugs that a moderately successful drug like a statin, the most widely promoted class of cholesterol-lowering drugs, used by one-quarter of the population over the age of forty, is touted as a solution. It should be acknowledged that statins can significantly lower the risk of a heart attack or stroke. In 2016, an article in the prestigious British medical journal *The Lancet* claimed that statins prevent 80,000 heart attacks and strokes a year in the United Kingdom. But what's hidden from sight is the difference between relative and absolute risks.

Let's say that a risk assessment conducted by your doctor told you that your risk of having a heart attack could be lowered 50 percent by taking a pill. That sounds impressive, but if your absolute risk of having a heart attack was only 10 percent to begin with, lowering it to 5 percent isn't so impressive. "Lowered by half" sounds dramatic compared with "lowered by 5 percent," which is why drug companies tend to give only the relative risk improvement. (For some, this relative risk reduction is essential. In fact, Rudy, who comes from a family with a strong history of early-onset heart disease, takes a statin as a necessary preventive in order to keep his LDL [low-density lipoprotein, or "bad," cholesterol] below 60. Family history can point to a genetic risk that would otherwise not be offset except by intervening with a cholesterol-lowering drug.)

Medical researchers present different perspectives on the same data. In a January 2009 letter to the *New England Journal of Medicine*, David H. Newman of St. Luke's–Roosevelt Hospital in New York City offers a striking example. The medical world had recently been encouraged by a meta-analysis showing considerable benefits from taking statins:

> The range [being cited] as the relative risk reduction for all-cause mortality, 20 to 30%, is inaccurate. The relative risk reduction was 12% over five years. This number means that one death per year was averted for every 417 patients taking a statin drug, or 1 of every 83 patients after five years of therapy. The large preponderance of subjects in this meta-analysis had established coronary disease, and the mortality rate among subjects in the control group (9.7%) was quite high.
>
> This benefit is real, but it is small—and substantially smaller among patients at lower risk (i.e., the great majority of patients currently receiving statin therapy). This benefit should be explicitly discussed with patients so that they understand, on the basis of their own five-year risk of death, their individual chances of benefiting from statin therapy, given the known risks and costs of these drugs.

Everyone, the public and physicians alike, welcomes good news about reducing the risk of heart attacks, and it's easy to forget the difference between relative and absolute risk. In absolute terms the risk reduction isn't large for patients who already have a diagnosis of CHD. Over a five-year period,

96 percent saw no benefit.

1.2 percent had their lifespan extended by being saved from a fatal heart attack.

2.6 percent were helped by preventing a repeat heart attack.

0.8 percent were helped by preventing a stroke.

0.6 percent were harmed by developing diabetes.

10 percent were harmed by muscle damage.

These findings are in keeping with general outcomes that find statins to reduce absolute risk in people with preexisting heart disease by an average of 3 percent, which is very different from the advertised lowering of relative risk, which was 20 percent.

The American Heart Association advises that taking statins is on balance useful. In support of this, a comprehensive 2016 review in *The Lancet* showed that statin therapy reduces the risk of major vascular events— for example, heart attacks and strokes—by incremental amounts that increase yearly if the drug is taken for five years. The overall estimate indicates that if 10,000 people lowered their LDL cholesterol with statins over this five-year period, 1,000 vascular events would be prevented; in other words, an absolute benefit of 10 percent. For those like Rudy with a strong family history that demands keeping their LDL levels low, this benefit is sufficient to justify taking the drug. But is it enough for people at lower risk? The current government recommendation is that after an assessment of risk with your doctor, you should take statins if your risk of heart disease is over 10 percent and you are between the ages of forty and seventy-five.

Even though they are accepted as the gold standard in cholesterol reduction, statins are by no means foolproof. Two published studies found that the calcification of plaques actually increased among statin users, making their heart disease progress faster. In one study of 6,600 men without previous diagnosis of CHD, published in the journal *Atherosclerosis*, the prevalence and extent of calcified plaque was 52 percent higher than in nonusers of the drug. Statins can also react with common blood pressure medications and blood thinners, along with antibiotics. Women of childbearing age taking statins must take birth control pills concurrently or else run the risk of the statins causing birth defects.

But let's set aside whether it's worthwhile to take statins for five years, counting the costs and the possible side effects (myalgia, or muscle pain, is common and increases with age or when taking other heart-related drugs). A more important statistic isn't understood by the general public: statins don't necessarily indicate that you will live longer. A meta-analysis by Dr. Kausik Ray and his colleagues published in *Archives of Internal Medicine* in 2010 found that statins had no effect on death rates from all causes. Statins work by managing one risk factor: they lower the levels of LDL cholesterol in the bloodstream, the low-density lipid considered to be the "bad" cholesterol. However, LDL levels were not found to significantly influence how long a person lives. One must take into account many other factors, such as inflammation and predisposition to calcification.

No doubt the cloud of risks for CHD is confusing, giving little indication which risk is key when you are making lifestyle choices. Is it cholesterol in your diet or stress at work? Is it sitting in front of the computer all day or carrying extra weight? Nor does the cloud of risks help out with another crucial factor—that as people age, entering the most dangerous decades for heart disease, they also tend to decrease their efforts to exercise, to follow a healthy diet, and to remain at an ideal weight. (A 2015 Gallup poll of 335,000 American adults reported that 51.6 percent say they do exercise at least three times a week for 30 minutes. But this doesn't really satisfy government recommendations, which advise 150 minutes of moderate to intense physical activity per week, plus two or more sessions of strength-building exercises that work all the major muscle groups. Only 20 percent of adults comply with this optimal amount, according to current Centers for Disease Control and Prevention data. People who tended to exercise the most were aged eighteen to twenty-six, earned more than $90,000 a year, lived in western states, and were male. Only two out of five obese people exercised at least three times a week.)

In a whole-system approach we want to clear away the cloud of risks and its confusion. To begin with, let's stop isolating the heart as if it were a vulnerable organ that has to be worried over constantly. The larger picture is very different. According to statistics from the United States and Europe, a person who is already sixty-five will on average live another nineteen to twenty years. The average holds true for both men and women but is strongly impacted if someone is poor, smokes, or is otherwise leading an unhealthy lifestyle. Yet if you ask a second question, how many of these extended years will be healthy, the answer is shocking: about half. A sixty-five-year-old man typically looks forward to eleven years of healthy life, a sixty-five-year-old woman slightly less. The term *healthy* is subject to varying definitions, but this overall picture is of a decade of diminished quality of life. Ultimately that's what we need to improve and prevent. A healing approach to the heart keeps in mind the bigger goal of wellness for life.

Heart Rate Variability (HRV)

We need firmer ground if we want to find healing that benefits the whole system. Let's begin with a measurement known as heart rate variability (HRV). The typical sound of a heartbeat is a steady drumming to the rhythm of a weak beat followed by a strong one: *lub-DUB, lub-DUB.* In reality, a healthy heart is flexible and changes its rhythm according to the situation. The pounding heart of a marathon runner is enormously different from the near cessation of the heart in the deep meditation of an Indian yogi. On a subtler level, your heart responds to the stimulation of daily stresses, even the most minor. If you grow tense, your heartbeat becomes more like a steady drumming, quick and evenly spaced. In medical terms, this means your HRV is low, which isn't preferable. In diabetics, low heart rate variability is associated with poor heart health and can even increase the risk of sudden cardiac death.

High HRV occurs when the heart responds within a flexible range of faster or slower beats depending on what's happening in the bodymind. On its own, a human heart beats at around 100 beats per minute, but the effect of the autonomic nervous system, which is responsible for unconscious processes throughout the body, reduces this to around 70 beats per minute. That's a desirable resting rate, on average. But it's ultimately the nervous system that proves critical.

When HRV is high, the autonomic nervous system is in balance. The signals that could cause fight-or-flight and stress reactions generally are kept in check by signals that promote rest and relaxation. When HRV is low, it can not only point to heart problems but also give other diagnostic clues about cancer, diabetes, stroke, glaucoma, and more. The far reach of these influences becomes intriguing to students of autonomic responses. You can intervene and slow down your heart by applying pressure to your eyes, for example, or by rubbing the carotid arteries on either side of your neck.

With the advent of wearable devices to monitor blood pressure, heart rate, and other vital signs, it turns out that HRV is one of the best single indicators of how stressed a person feels. Using simple deep breathing or a few moments of meditation, people can improve their HRV while reducing the stress response. A wearable device can monitor and verify the change. Thus, subjective and objective reality merge together, as they already do in the union of body and mind.

Let's say you're late for work and rush out of the house. It's a cold morning, and when you turn on the ignition, your car won't start. At that moment, both sides of reality begin to have an effect. On the objective side is an outside stressor—your car's dead battery—which leads to objective changes in your body. Stress hormones like adrenaline and cortisol are likely to kick in; the brain's emotional center, the amygdala, will show heightened activity; your blood pressure may rise, and your heart rate increase. All of this is typical of the body's stress response. On the subjective side, the range of reactions is so variable that they are far less predictable. You could panic, for example, if this is your second week on the job and getting fired would be a disaster. On the other hand, you could own the company and consider this a minor inconvenience. Outside the realm of small daily stresses, the major stresses that alter a person's life evoke everything from grief and sorrow to extreme fear, depression, and suicidal tendencies.

The wonder is that the whole system is so sensitive and dynamic that it handles the entire spectrum. But at bottom, the pivotal factor is subjective. How you perceive and interpret any stress determines how hard it affects you. A dead car battery can be the start of something big or nothing at all. So how do you deal with a measurement that depends so heavily on your inner life? It's an important question, because in terms of risk factors, low HRV is linked to a number of disorders, both psychological and physical. As a marker for mental illness, low HRV appears more or less everywhere, from depression and generalized anxiety disorder to post-traumatic stress disorder (PTSD), bipolar disorder, and schizophrenia. The heart suffers the mind's distress. On the physical side, low HRV is associated with inflammation, opening the gateway for a range of disorders so wide that, once again, low HRV may be a marker for illnesses that seem to have nothing to do with one another, like cancer, diabetes, and heart disease.

Clearly it's good to have enhanced HRV—that much is medically certain. A direct way to achieve this is through meditation and other contemplative practices, as we've already mentioned. If you look back at the illustration of the bodymind pathways on page 15, you'll notice that the heart is placed midway between the "bottom-up" messages sent from the intestinal region and the "top-down" messages sent by the brain. To a physiologist who is looking for a specific part of the anatomy responsible for routing messages in both directions, what stands out is the *vagus nerve*, so let's look at it more closely.

STIMULATING THE VAGUS NERVE

The term *vagus* derives from the Latin for "wandering," which is what the vagus nerve does. It is one of the ten cranial nerves that branch directly from the brain to spread out across the body. The vagus nerve wanders from the brain to the intestines, with stops along the way, primarily at the heart and lungs. Its major responsibility is to regulate heart, lung, and digestive functions. As the longest nerve in the body, the vagus has two main branches, right and left, that descend down either side of your neck. The vagus nerve can be considered hardwired from the gut to the brain. The signals are produced by the gut microbiome, the bacteria that inhabit the gut. The gut microbiome contains 200 times more genes than the human genome (4,000,000 versus 20,000).

Of the many facts that can be stated about nerves, one important distinction is that some nerves send signals away from the brain (efferent nerves) while others send signals back to the brain (afferent nerves). Because of its countless small branches reaching almost every organ, the vagus nerve is responsible for between 80 percent and 90 percent of afferent impulses. In everyday language, this means that the sensory information—especially the effects of pain and stress—traveling along the body's information superhighway travels along this one nerve. As a result, when vagal nerve activity is low, a host of things could be going wrong—diminished activity is associated with increased death from infections, rheumatoid arthritis, lupus, irritable bowel syndrome, sarcoidosis (a disorder of unknown cause that creates swollen lymph nodes), trauma, depression, and stress. Stimulating the vagus nerve has an instantaneous effect on your heartbeat and HRV.

By now you've become used to these laundry lists of illnesses that cross the bodymind frontier. Here, it's significant that the vagus nerve is a two-way highway, sending signals back and forth between the intestinal tract and the brain. It regulates the gut-brain response, which in turn may be all-important in inflammation. Published findings indicate that meditation and various contemplative practices may improve immune response by stimulating the vagus nerve in such a way that inflammation is reduced.

One enticing piece of evidence involves stimulating the vagus nerve physically. This involves the surgical implant of a small battery-powered generator about the size of a watch. Typically done as an outpatient procedure, the implant is tucked into the space below the left collarbone. A wire is run into the neck where the left branch of the vagus nerve descends, and the wire is looped around the nerve. When the generator is turned on—there are various settings from weak to strong—it sends faint electrical pulses that stimulate the vagus nerve.

What's mind bending, as viewed from conventional medical training, is how wide the possible benefits of vagus nerve stimulation (VNS) seem to be. Presently no less than thirty-two disorders are undergoing research, with indications of positive results. They begin with alcohol addiction, irregular heartbeat (atrial fibrillation), and autism, running through a rogue's gallery of physical and psychological illnesses: heart disease, mood disorders like depression and anxiety, a variety of intestinal disorders, addictions, and perhaps even memory loss and Alzheimer's disease. The "wandering" nerve embraces many parts of the brain and body, implying that it can have a healing effect holistically.

A breakthrough finding about the vagus nerve came from Kevin J. Tracey, a neurosurgeon and a specialist in molecular medicine. Dr. Tracey had the insight that the body's immune system must have evolved to preserve homeostasis, the overall balance of the body. When inflammation occurs as a part of a normal immune response (the so-called inflammation reflex), the body goes out of balance, entering healing mode. There are specific chemicals that regulate this reflex, which is controlled from the nucleus of cells. One important inflammation marker is a chemical group known as cytokines. It's known that cytokines can go outside their normal range, and when this happens, acute or chronic inflammation occurs. By analogy, the body lights a fire that won't go out and can flare up dangerously.

It was long assumed that the immune system took care of the inflammation reflex on its own, but beginning in 2011 Tracey and his colleagues showed that a brain chemical, or neurotransmitter, called acetylcholine is engaged in regulating how much or little cytokine is produced. Specifically, they connected acetylcholine with memory T-cells in the spleen. The highway that this messaging travels along is the vagus nerve. (A May 2014 profile of Tracey in the Sunday magazine of the *New York Times* was aptly titled "Can the Immune System Be Hacked?")

Tracey and his team went on in 2012 to demonstrate the therapeutic benefit of stimulating the vagus nerve in a paper that disclosed improvements in rheumatoid arthritis symptoms that were resistant to conventional drug treatment. This result opened the floodgates of research in many areas. Suddenly the brain-gut axis became one of the hottest topics in internal medicine. The paradigm for diseases is now undergoing radical revision, and the revision is always in the direction of wholeness, viewing the bodymind as a single system.

For example, millions of people suffer from irritable bowel syndrome, also called by related names such as spastic colon, nervous colon, and mucous colitis. It's a miserable illness, causing not only severe abdominal pain and irregular bowel movements but also the psychological distress of never knowing when these unpredictable symptoms will strike. Looked at locally as purely an intestinal problem, an irritable bowel displays inflammation that makes the intestinal area hypersensitive—a minor stimulus can lead to raging inflammation.

Now the paradigm for the disease has changed with the discovery that, via the vagus nerve, different brain areas are involved, such as the somatosensory cortex, insula, amygdala, anterior cingulate cortex, and hippocampus. Both outgoing and incoming nerve signals are involved along the brain-gut axis, and once the brain is invoked, there's a major opening for emotions and stress reactions. This helps explain why many sufferers from irritable bowel syndrome find themselves referred to psychotherapy, because the dysfunction in their everyday routine makes them so anxious and depressed.

However, now that we know that there is abnormal brain signaling in patients with irritable bowels, separating the physical and psychological aspects of their illness no longer makes sense. One of the main causes for optimism in treating the disorder is the vagal nerve implant, because it would enhance brain-gut axis activity. Also on the horizon are wearables that stimulate the vagus nerve without surgery, sending the faint electrical pulses through the skin where nerves are closely connected to the vagus nerve, such as around the ear.

The vagus nerve is a convincing example that supports the idea that there is only one bodymind—in effect, it's a lifeline to the heart carrying messages of both physical and mental events. A completely noninvasive alternative is meditation, long proved to reduce the stress response; there are anecdotal reports of meditators saying that their irritable bowels became better after they began meditating. The breakthrough discovery that the brain directly affects T-cells in the immune system was startling, even to experts in the field. Medical school training always kept the central nervous system in one compartment and the immune system in another. Now it is suspected that there are dozens of lifestyle choices that can stimulate the vagus nerve. This puts in place a critical piece of the jigsaw puzzle, showing that the immune system is connected to the brain and not an isolated system on its own. But all of this physical evidence must not pull us in the wrong direction. Healing is not only controlled physically; the key is consciousness.

Awareness and Inflammation

We keep repeating that what you aren't aware of, you can't change. One key link—perhaps the ultimate answer—to many chronic disorders, including CHD, seems to be something it's difficult or impossible to be aware of: inflammation. The initial effects are microscopic in the case of what's happening inside the heart. This requires a bit of medical explanation. The slick inside surface of your arteries is known as the endothelium. Its slickness isn't entirely because of having a smooth surface. The endothelium is dynamic and active. It secretes chemicals, for example, that repel toxins that are damaging to it, like the residues of smoking. Unlike water pipes, blood vessels expand and contract to alter how much blood flows through them. The rigidity of the plaque that builds up in heart disease is a problem, but the underlying issue is atherosclerosis, commonly called hardening of the arteries.

Just as leaves gather in a rain gutter in the fall, when the endothelial lining of the coronary arteries starts to develop cracks, bits of LDL cholesterol floating by get lodged in them, and the fatty deposits are slowly hardened with calcium buildup and tiny blood clots. Over time, the white blood cells that rush in to deal with the LDL cholesterol inside the artery wall also add to the plaque. (Atherosclerosis isn't limited to the coronary arteries but is a systemic disease. The potential for strokes is often associated with plaques in the carotid artery in the neck.) It's known that high blood pressure, smoking, and high LDL levels are causes of plaque.

But they aren't how the disease begins. At the microscopic level, the first signs of atherosclerosis seem to occur as fatty streaks in the artery's muscle cells. These fatty streaks apparently become inflamed, and from there the cracks in the endothelial lining develop. No one knows where the fatty streaks originate, but it seems likely that by the time traditional prevention is followed, the disease is well past its beginnings. Yet in between the fatty streaks and the cracks, one culprit—inflammation—can be attacked. In fact, from everything we know, this appears to be the best whole-system approach.

Inflammation is a whole-system problem from head to toe. But if we can't detect it in everyday life, how do we know what to do? Unlike the redness, swelling, and discomfort of *acute inflammation* (e.g., at a burn or wound site), low-level *chronic inflammation* causes few, if any, symptoms. Telltale markers for inflammation, particularly cytokines, show up in artery walls afflicted with atherosclerosis. Where awareness can make an important difference is in the area of stress, which, as is well documented, creates inflammation.

Meditation reduces stress by acting at the level of unconscious autonomic responses in the brain. But as you become more self-aware, either through meditation or some other avenue, you begin to pick up on negative input that keeps the whole system in a low-lying stress response. The chemical story is complex, but the chain of damaging events is clear:

Stress ⟶ Inflammation ⟶ Atherosclerosis ⟶ CHD

If the first link in this chain of events is instead *self-awareness*, the rest of the sequence is either preventable or reduced, making treatment easier. Earlier (see page 41), we gave some pointers on how to remain self-aware at work. But self-awareness can be blocked by all kinds of things. Using the workplace as an example:

- The pressure of deadlines induces a level of stress that's chronic, and we adapt to it by blocking it out, eventually even normalizing it.

But our cells don't have this blocking mechanism and incrementally become damaged.

- Heart rate variability suffers under the constant demands of a typical workday.

- The sedentary routine of modern jobs, many of which involve hours at the computer, weakens muscle tone and adds to the current epidemic of obesity.

- The repetition involved in routine work dulls the mind and creates a flattened mood.

- Interpersonal tensions in the workplace create resentment, anger, envy, and anxiety that wind up being shoved down out of sight without actually being resolved.

- Unexpressed negative emotions and tension are communicated back and forth between the brain, heart, and gut along the vagus nerve, creating diminished function that often shows up in a tight stomach, an irritated bowel, constipation, and other signs of inflammation.

These workplace stressors offer a prime example of how "normal" life is actually working against healing, and similar stressors exist outside the workplace, at home. However, no matter how slow the pace of dysfunction is, the whole system is paying a steady price that mounts up, bit by bit, every day. When you go to work, you are taking 50 trillion cells to the office with you, and *their* well-being determines *your* well-being in the end.

Inflammation is a complex issue that largely occurs at a hidden cellular level, but the stress response is something we can control in everyday life. Ironically, this is the element most people pay the least attention to. They improve their lifestyle with diet and exercise while leading fast-paced, demanding lives that lie at the heart of the problem. The next step in our healing journey is to look at how stress and the healing response are connected at the deepest level.

5

Getting Out of Overdrive

Decades after stress *became a household word, most people don't* understand it—and that's not their fault. Ask yourself which of the following events you would consider stressful:

Going through a divorce

Winning the lottery

Going on vacation

Having a baby

The correct answer is "all of the above." *Stress* can be defined as any input that triggers the body's stress response. Psychologically, stress results whenever anxiety about the future or regrets about the past are invoked. We may label a nasty divorce as a negative event and winning the lottery as a positive one, but that's not how the lower brain sees it. The lower, or reptilian, brain is an evolutionary inheritance from earlier stages of life on Earth, bringing with it our ancestral fight-or-flight response. A vast range of everyday experiences, from giving birth to losing your job, from having depression in your family history to breaking the bank at Las Vegas, can be stressful. Experts refer to "eustress," indicating that the stressor is a happy event—the Greek prefix *eu* means "good" or "well"— but it is stressful nonetheless.

In a healing lifestyle we must address the stress that accumulates over time, and because most of this accumulation is invisible and occurs very slowly, there is really no difference between stress management and life management. For example, joyous as it is to give birth, new mothers report that raising an infant is extremely stressful—not that this should come as a surprise. We adapt to eustress and distress, the good and bad along the spectrum, because we must. Babies need to be cherished and nurtured, despite the cost in both physical and mental stress to the parents.

A whole-system approach tells us that putting up with stress isn't good enough. New parents are assured by everyone who has been there that babies eventually stop waking you up in the middle of the night, stop teething, stop being terrible two-year-olds. All true, but there is always something stressful lying ahead, which is true of life in general. So dealing with stress involves two things: clearing from the system the residue of old stressors and preventing the impact of new stressors from being too severe. Both steps figure prominently in life management.

Handling Acute Stress

Some events strike out of the blue and face us with immediate stress—these are known as *acute stresses*. Being fired is an example, and all of us have experienced how difficult losing a job can be; it's something deeply feared by millions of people. We have also experienced self-defeating ways of dealing with this kind of crisis. A certain percentage of people simply withdraw and seek distraction, hoping that time will heal their wound. Psychiatric studies have found, for example, that the most common behavior in the face of acute stress is watching more television (which now would be updated to constantly playing video games), a phenomenon that has become endemic among older blue-collar workers laid off from work, often permanently. Since this behavior is also attended by higher rates of opiate addiction among white males over fifty and an alarming increase in suicides, using distraction clearly isn't a viable defense against acute stress.

In your own life, when you are hit by acute stress, such as a bad breakup or the diagnosis of a potentially serious disorder, a certain amount of withdrawal and distraction is natural and good. Time doesn't lead to complete healing, but it allows disrupted emotions to return to equilibrium. Turning to comfort food and "eating your feelings" make sense emotionally for a while. But eventually you need to cope with acute stress in a healing way that is proactive. Otherwise you may walk away with lasting wounds, bad memories, lower self-esteem, and other damage.

The path to healing is actually demonstrated by the situation of having a baby. After the mother gives birth, her brain creates higher levels of dopamine and oxytocin, two chemicals associated with heightened mood, even euphoria. As with any pleasure or reward a person experiences, the first time makes us want it again. A 2008 study headed by Lane Strathearn at Baylor University and published in the journal *Pediatrics* showed that when new mothers experienced pleasure upon seeing their baby, the same brain regions were activated as those activated by cocaine—in this case a natural high. Intriguingly, seeing either a happy or sad baby face was pleasurable, as measured by reward signals in the mother's brain, as long as she felt confident and secure about her baby. In contrast, mothers who were overly stressed about their newborns had different areas of the brain activated when the baby cried, areas linked to pain and disgust. It turns out that the stress level of a mother can have a dramatic effect on her interaction with the infant and how the baby's own brain will develop.

The stress that occurs when a newborn arrives isn't simply going to vanish. For a year or more, both parents live disrupted lives, and the typical signs of acute stress rise up to greet them, including fatigue, frayed tempers, bad sleep, and a sense of losing control. Stress experts indicate that increasing the unpredictability in your life and feeling that you are out of control makes acute stress much worse. It's easy to see how losing your job turns a steady income and pride in doing a good job into the opposite, having no accomplishment to feel proud of and not knowing what the future will bring. But having a baby has the same dimensions. Infant health is unpredictable, and parents have no control over when their baby is suddenly going to need immediate attention.

Some parents of newborns do much better at coping than others; see below for examples of how they do it.

The "Baby Solution" to Acute Stress

The key is a range of coping mechanisms anyone can call upon.

- Get enough rest and sleep.
- Make time for yourself every day to be alone and quiet.
- Make sure you get outside to refresh your connection to nature.
- Maintain an active life—don't be chained to the situation.
- Share duties and responsibilities. Ask for help before you feel over-whelmed.
- Pursue a regular routine—this helps offset unpredictable events.
- Find an activity that makes you feel in control.
- Find a confidant with whom you can share your feelings without judgment.
- Don't martyr yourself by taking on more than you can handle.
- Fight the urge to feel victimized.
- Don't isolate yourself—keep up your social activity.
- Seek out people in the same situation who can empathize with you and offer positive support.
- Resist self-judgment. Be easy on yourself, accepting the ups and downs of emotions as natural.
- Where there is the possibility of finding joy, pause to appreciate it.

The arrival of a baby is such a joyful event that the positive side that counters the stress is obvious and easily accessible. The same isn't true if you are going through a divorce or suddenly lose your job. Even so, the important point is to be aware that you can cope by developing the coping behavior we've outlined. This is a conscious, self-aware project. Your built-in responses can't accomplish it for you.

If you find yourself in a crisis that induces acute stress, take the following steps:

1. Start to journal about your path out of the crisis.
2. In your journal write down the list of coping mechanisms we've just given. You might want to make each one a page heading on its own.

3. Under each coping behavior, write down something you can do immediately to adopt that behavior.

4. Follow up every day with your successes when a coping mechanism is beginning to work for you.

None of these coping mechanisms is complicated; most are self-explanatory. But acute stress is such a powerful disruption that it throws our awareness off kilter. We wind up doing things we know deep down are self-defeating, such as being alone too much of the time, acting out the role of a victim, and letting fear and anxiety gain the upper hand by keeping our emotions bottled up.

We've already described how people who feel supported are significantly less likely to develop angina than those who don't feel such support (see page 39). The connection between emotional well-being and heart health is undeniable. But the same truth holds for confronting acute stress, which threatens health and well-being at every level, including your body. Yet in most lives, situations that create acute stress will tend to be intermittent and hopefully rare. We need to extend the discussion to the invisible kind of everyday stress that actually causes more harm than people realize, harm that can prove disastrous over a period of years without being detected. We're referring to the hidden culprit, *chronic stress*.

Chronic Stress and "Sympathetic Overdrive"

How are you handling the little everyday stresses in your life? Most people complain about the stressors that plague almost everyone in modern society, particularly its high speed, long work hours, and inescapable irritations like traffic jams and boring commutes. Our inclination is to adapt to these stresses, taking them in stride. We shrug off the way life keeps speeding up (even demanding more speed when it comes to the Internet and smartphones); we listen to music to divert the frustration of traffic snarls and long waits at the airport; we accept that work pressures are necessary in order to get ahead in our careers.

Human adaptation is a miracle, but stress management took a wrong turn at the outset when experts, and the medical profession in general, focused on two factors as the most important: physical stress and outside stress. These go hand in hand. The theory was that some external event triggered a physical response in the body, and in this interaction the main problem of stress was revealed. So if you hear gunfire (outside stressor) and instantly feel your pulse racing (physical response), the typical stress reaction has been triggered. True, this pattern is common enough. We mentioned earlier a string of outside events that are highly stressful, such as going through a divorce or winning the lottery.

But from a whole-system viewpoint, at least half the story is untold, because the inner world of subjective events creates stress, too, and at the same time is the source for healing the effects of stress. Let's look at a highly stressful event, entering the hospital for surgery. On the physical side, the stressful event is the medical procedure itself, but other stresses are having an impact mentally and emotionally. These include the following:

Worrying about the surgery's outcome

High or low expectations

Trust or distrust in medical care

The strangeness of the hospital environment

Disruption to normal daily habits

Intrusive and embarrassing poking and prodding

Loss of control over what is happening

Anxiety about the future

Fear of what will happen to the whole family

So much depends on these factors that they should be placed first and foremost. To a surgeon, either he successfully repairs a diseased heart, liver, or brain or he doesn't. But the physical outcome barely touches upon the invisible stresses and how we handle them.

Because the inner approach to stress, including meditation and mindfulness practices, has proved to be so beneficial in reducing stress, it's reasonable to think that going inward is how the average person fights against stress. But it would be an exaggeration to believe that meditation and mindfulness have penetrated deeply into typical Western lifestyles. Why? There has been plenty of media coverage about meditation and its benefits. The negative attitudes toward it have steadily melted away—few people now consider meditation a strange esoteric religious practice from the East. The resistance to adopting meditation and mindfulness gives us a portrait of life that is stuck in old habits and attitudes that not only block meditation but also block a healing lifestyle generally.

Without seeing the damage we are doing, most of us, the vast majority in fact, have put ourselves in overdrive. What does this mean? In physiological terms, the best reference is the nervous system. The nervous system is the prime example of how your body operates under dual control, a point we keep returning to. Any process that you don't have to think about is handled by the autonomic nervous system, which in lay terms was once referred to (somewhat mistakenly) as the involuntary nervous system. Essentially the autonomic nervous system controls how organs function. The term *involuntary* once made perfect sense, because the nerves that control the heart, stomach, and digestive tract oversee functions that don't need our voluntary cooperation. You can't tell your heart to stop beating or your small intestine to give you a break and extract fewer calories from the food you eat.

But the idea that you have no control over the autonomic system is misleading, because this nervous system turns out to be more adaptable to our wishes, feelings, thoughts, and other mental activity than anyone used to think. The autonomic nervous system is divided into two parts, known as the *sympathetic* and *parasympathetic nervous system*. (Again the terms are a bit misleading, since *sympathetic* isn't being used in the same way as extending your sympathy to someone else.) The basic function of the sympathetic nervous system is to deliver the fight-or-flight response. Even though the lower brain is the seat of fight-or-flight, it takes a whole network of nerves throughout the body, extending from the spinal cord, to activate everything that goes into this one response.

A great many elements are involved in the fight-or-flight response: pupil dilation, increased sweating, increased heart rate, and increased blood pressure. At the same time, digestion is temporarily halted, metabolism shifts into another gear, and muscles begin to operate anaerobically—that is, without needing oxygen. Being temporary, these are emergency measures only. Evolution didn't equip us to react to stress constantly. In addition, when a full-blown stress response is triggered, there is little one can do to override it, because the hormones being secreted, such as cortisol and adrenaline, latch on to specific receptors on the cell membrane and trigger a chain of unstoppable events inside the cell. For example, in bone marrow, chronic stress can cause immune cells to promote inflammation, a process that begins with changes at the level of genes. If a certain stressor, like a loud neighbor, is happening every day, chronic inflammation can be the result leading to heart disease, cancer, or other disorders. Luckily, these detrimental changes in our cells in response to stress can also be temporary. From this picture, instead of saying that most people are in overdrive, what we should call it is "sympathetic overdrive," because too much demand is being placed on the sympathetic nervous system. Fight-or-flight feels like an on-off mechanism when you experience it: the signs are drastic and unmistakable.

If you've ever seen street magic on TV or in person, when the magician performs a trick, whether it's pulling the ace of spades from behind someone's ear or correctly reading a random number they are thinking, many spectators actually run away—they may be laughing, but the sympathetic nervous system can't take a joke and forces them to flee, at least for a moment.

Yet, in reality, the stress response operates on a sliding scale, and the sympathetic nervous system can be thrown into a low-level state that over time produces a wide range of damaging effects.

Beyond what most people realize, being in sympathetic overdrive undermines them every day. We can illustrate the problem with a story about a woman we'll call Mara, whose life contains nothing disastrous or deeply troubling but who exemplifies how far from healing many people actually are without knowing it.

Mara's Story: Invisible Damage Over Time

Mara is forty, successful, and has little to complain about. She learned early on that she was a bright student, and academic achievement followed her through school until she graduated with honors from an Ivy

League college. This was in the mid-1990s, and like many other young people buoyed by a booming economy, she went into the financial sector, getting a good entry job at a major bank. Her life began to unfold according to plan.

"I made very good money, and I got promoted on the fast track," Mara recalls. "The price was total dedication to the job, and like everyone I knew, I spent a minimum of sixty hours at work every week. I took work home and sometimes came into the office on Saturdays. Frankly, I enjoyed it. When I heard that some people thrive on stress, I thought to myself, *This is me*."

Mara developed this attitude after discovering very quickly how competitive her chosen career actually was. Her friendships soon became limited to coworkers at the bank—young, ambitious types she found exciting to be around. They were determined to be winners. Out of this pool of people she began to date Frank, another banker, who was also attending law school at night.

"Frank was driven," Mara says, "but he was also smart and funny. He could size people up and cut them down to size if he had to. It seemed like we made a great team."

With matching lifestyles and goals, they became a serious couple and moved in together. With such a strong focus on work, they decided to postpone having a baby at least until their thirties.

Fast-forward five years. At thirty Mara had moved on to a new relationship—looking back at the three years she spent with Frank, they were probably too much alike. They both had strong egos; they argued a lot, and neither liked to back down. Yet what finally broke up the relationship was money. When Mara began to make more than Frank did, he sulked and tried to compensate by trying to act more dominant and aggressive, looking for excuses to put her down.

"I wasn't that shaken up when he decided to move out," says Mara. "I suspected he was looking around for somebody else anyway. I bounced back pretty quickly, and it was only a few months later that I met Jason, who wasn't ambitious and had a career totally outside finance. Jason is tender, caring, and noncompetitive where Frank was self-centered, tense, and angry. Once I saw the contrast, it was easy to make a change."

Mara's career was still moving ahead, but she did notice that males at her level were getting promoted ahead of her. This and other evidence of sexism made work more problematic, but she was good at her job. She also started paying attention to exercise—she jogged regularly now—and watching her weight, two things that hadn't really been part of her lifestyle in her twenties.

Jump ahead to age forty. Mara married Jason and they have a four-year-old girl. After taking three months of maternity leave when the baby was born, Mara was back at work. She feels good about her relationship with Jason, but there are areas of conflict. In particular, he revealed himself to be passive and sometimes does things she considers passive-aggressive, like "forgetting" to pick their child up from day care on the day after a big fight with Mara. In the dynamic of their marriage, Mara has adopted the aggressor role, although she hates being that way, while Jason goes quiet and watches TV if he senses tension, even though she's begged him a thousand times to tell her how he really feels.

"I look around, and things aren't perfect," she says. "I sort of fell into the supermom syndrome trying to be a total achiever at work and a caring wife and mother at home. It's going okay. There are plenty of people worse off than me."

There are other positive aspects of her life that she rarely even thinks about. Mara's health is good, essentially as worry-free as when she was twenty. She's never had a cancer scare, and being premenopausal, her body's estrogen has protected her so far from heart disease. It's true that she stopped jogging during her pregnancy and never returned to it, and she intermittently attempts to diet to lose the ten pounds she gained back then. But with increasing emotional maturity, she is better able to navigate the ups and downs of an intimate relationship as well as raise a child like a responsible, loving parent.

So where's the problem? Millions of people lead similar lives and feel that nothing troublesome is affecting them. Yet if you consider what we've discovered so far about a healing lifestyle, Mara isn't living it. See below to gain insight as to where the invisible cracks can occur in the life you are likely to be leading now.

How a "Normal" Lifestyle Blocks Healing

- Daily activity gets driven by work, with its demand for achievement and success along with our fear of loss and failure.

- Self-esteem is built on external norms like getting promoted and being competitive.

- With so much focus on externals, life is lived on the surface. As the outside factors get more organized, a person's inner life doesn't keep up.

- Emotional needs are placed second or not faced honestly.

- Little or no attention is paid to chronic low-level stress.

- Relationships settle into routine and habit.

- Physical activity and contact with nature begin to diminish over time. Life becomes gradually more sedentary.

- There's no higher vision of possibilities, thanks to the burden of constant demands and duties from family and work.

- Paying attention to health issues is temporary and intermittent. For the most part, little is done until actual symptoms appear.

This is a shocking list of things we take for granted—or manage to put up with—even though they keep us in sympathetic overdrive. Stress follows in the wake of each item on the list, which means that stress is a much bigger issue than we assume. To put it simply, millions of people put a positive value on choices that are actually negative from a whole-system perspective.

So where do you stand right now? It's hard to assess your own stress simply because the bodymind system is so good at adapting. Years go by without seeming to expose the damage stress is doing. Stress experts recognize three stages of stress that occur one after the other. The earliest stage exposes psychological effects; the next stage, behavioral effects; and the third stage, physical effects. What follows is a summary of each. Read each category to see if you detect any signs of stress taking its toll on you.

Three Levels of Damage

Psychological and Neural

Psychological and neural damage begins with minor things like feeling mentally tired and under pressure from deadlines at work. When people say they are stressed out, they generally mean that they've run out of energy, which can mask mental states like being depressed, anxious, or even panicky. Because the brain is being affected, there's interruption in normal sleep rhythms or the nagging feeling that time is running out, a condition Dr. Larry Dossey dubbed "time sickness." With mental fatigue can come flawed decision making or memory lapses, but in general the problem is loss of concentration, the ability to focus. Emotionally, stress seems to make us retreat to infancy, becoming prone to outbursts of anger, distress, and irritability. The more the stress mounts, the shorter the fuse gets on our negative emotions.

Behavioral

Negative changes in behavior are likely to manifest in two major areas, work and relationships. Stressful jobs make us respond with all kinds of behaviors, from office gossip to going out for a drink after work. As stress mounts, the drinking can get heavier, the need for distraction more severe. Inevitably we take our feelings home after work, where friction easily follows. A spouse who feels neglected, mistreated, or ignored is feeling the brunt of stress-related behavior. Stress makes one person lose his appetite and another overeat. Sleep often gets disrupted, and in some cases chronic insomnia is the outcome. This and other ill effects might lead the person to become dependent on sleeping pills and other drugs in an attempt to shake off work stress and find the way back to feeling normal.

Physical

When the body can't completely adapt to stress, bad effects follow without being predictable. Most people will suffer from physical fatigue. Stomachaches, bad digestion, and headaches are likely. So is a reduced immune response, leading to more colds and worsened allergies. After that, the problems will tend to be associated with inflammation, whose effects can travel anywhere. One person may experience skin eruptions, another irritable bowel syndrome, yet another a heart attack or stroke. By this stage, the damage caused by stress has led to serious system breakdown.

In the brain, stress activates a specific neural network called the *hypothalamus-pituitary-adrenal (HPA) axis*. Activation of the HPA axis leads to the excess production of specific hormones known as glucocorticoids by the adrenal glands. Glucocorticoids are needed for normal brain development and are also activated during moments of acute stress. However, elevated levels can have the reverse effect and cause neurotoxicity, as shown in studies of stress during pregnancy. There is a natural barrier that prevents stress hormones in the mother from being passed through the placenta to the fetus. In stressful pregnancy, this barrier appears to be crossed, and one major result is interference with normal brain development—when glucocorticoids were given to pregnant rats, the brains of the offspring didn't develop properly.

Much more than was suspected in the past, a difficult pregnancy that puts the mother under chronic stress can have far-reaching effects at the cellular and genetic level. In humans, excess glucocorticoids in the prenatal brain of the fetus directly affect levels of dopamine, which, as we saw earlier, is involved in reward or pleasure seeking. Prenatal stress may also have downstream effects as infants grow up, including learning disabilities, higher susceptibility to drug abuse, and increased anxiety and depression. Maternal stress has also been linked with increased HPA axis activity at different ages of the child, including at six months, five years, and ten years, and onward to adulthood. Disturbingly, in animal studies these elevated levels of glucocorticoids linger genetically into the next generation or two.

We don't offer any of this information to alarm you, only to show that low-level stress deserves to be called the epidemic of civilization. With its tentacles reaching everywhere, no one is immune. The dilemma is that, being so pervasive, chronic stress causes too many things to potentially go wrong: experts have found no single remedy that can cope with the unpredictable outcomes brought about by everyday stress. Let's see how a healing lifestyle can do better.

The Whole-System Answer

You won't be surprised that the whole-system answer is to bring awareness into the picture. Since the first ill effects of chronic stress are psychological and neural, this is where healing also begins. We've already mentioned that putting up with stress and adapting yourself to it are bad strategies. Your cells aren't adapting even when you think *you* are. A good example would be workers on the night shift. Long stretches of night work disrupt the body's circadian (or daily) biorhythms. As a result, the most obvious detriment is the loss of good sleep, which has long been known—the brain never fully adjusts to a schedule of not sleeping at night. But further investigations have revealed that night-shift workers are at risk in seven other ways:

Higher risk of diabetes

Increased likelihood of obesity because of hormonal imbalance affecting hunger and satiation

Increased risk of breast cancer

Negative metabolic changes that could influence heart disease risk

Potential increase in heart attacks

Higher likelihood of workplace accidents

Higher risk for depression

In short, the whole system is potentially affected due to putting too much disruption in a single biorhythm that turns out to be connected to other biorhythms, like the link between sleep and hunger-satiation. It also seems that the obvious solution—quitting your night-shift job—may not be enough to reverse the damage if someone has been working too many years at night.

The basic lesson for everyone is that stressors aren't isolated things. A blanket behavior or attitude can spread its bad influence very widely. Let's say you're at the airport and find out that your flight has been canceled. The airline won't bring another airplane into service but tells you that you must wait five hours until a flight arrives that can accommodate you. With no alternative except to comply with the airline's mistreatment, passengers look passive as they sit and wait, but on the inside many people (perhaps you) will react with the following responses: worry, complaining, and pessimism. All are self-defeating.

Worry is self-induced anxiety. It solves nothing and blocks the possibility of dealing with things more positively.

Complaining increases tension and anger. As a display of hostility, it encourages other people to act hostile in return.

Pessimism induces the illusion that a situation is hopeless. It fosters the belief that expecting a bad outcome is always realistic, when in fact it isn't.

If you see yourself in any of these behaviors and attitudes, you are fooling yourself into believing that you are adapting to stress. As your body experiences it, however, you have become the stressor yourself. That's because an external event (canceled flight) must go through an internal interpretation before it triggers the stress response. Unlike a crisis like losing your job, a flight delay belongs in the category of everyday chronic stresses. Which means that you have a choice to respond. Worry, complaining, and pessimism are unconscious responses. People who are stuck in them have become the victims of old reactions that became glued in place because the person didn't reevaluate them.

Some people handle a canceled flight better than others. Just as we gave you the "baby solution" for acute stress, here's the "airport solution" for low-level everyday stress.

The "Airport Solution" to Chronic Stress

Detach yourself from the stressor. At the airport people do this by reading a book or finding a place to be alone.

Become centered. At the airport people do this when they shut their eyes to meditate.

Remain active. At the airport this means walking around instead of slumping in a chair and waiting.

Seek positive outlets. At the airport this might mean shopping, getting a chair massage, or going to a restaurant.

Rely on emotional support. At the airport the usual way to do this is by calling a friend or family member on the phone. (A short call announcing that you'll be late won't give you emotional support. The key is a conversation with someone meaningful in your life that lasts at least half an hour.)

Escape if you must. At the airport, if the airline's behavior gets too outrageous, it saves your psychology to reschedule and go home. (Of course this is not always practical or affordable.)

All of these things are positive adaptations, as opposed to the negativity of worry, complaining, and pessimism. They bring awareness into a situation where falling back on passive acceptance isn't the right answer. Beneath the attitude of "I have to put up with it" lies stress. A canceled flight is usually not fixable by you, and it can happen anytime without warning. Therefore, it fits the two conditions that make stress worse: unpredictability and loss of control.

You have the option of turning the situation around by interpreting it not as bad luck but as a non-stress, to which you respond by doing things you actually want to do, like meditating, connecting with a friend, or shopping. When you become adept at this turnaround, chronic stress is nipped in the bud. You cut short a process that otherwise would have affected your body like Chinese water torture, drip by drip.

The "airport solution" applies right now as well. It describes a strategy for getting yourself out of sympathetic overdrive. There's a physiological explanation for what happens. The sympathetic nervous system is balanced by an entirely separate set of nerves with opposing responses known as the parasympathetic nervous system. Instead of tension, it brings relaxation. As nature designed them, the sympathetic and parasympathetic nervous systems are antagonists, we can say. The temporary, drastic action of the sympathetic nervous system is countered by the continuous, balancing activity of the parasympathetic nervous system.

Under chronic stress, we ask the sympathetic nervous system to stay on guard all the time, until it gets out of its normal groove and begins to impair the normal state of balance. At the same time, the relaxed, normal state of the parasympathetic nervous system gets blocked or sidelined. To get out of overdrive, you must enhance the parasympathetic side of the equation. This can only be done by applying conscious choice, since left to its own devices, the two antagonists will keep doing what they are habituated to do. Absent the influence of stress, the automatic back-and-forth of sympathetic and parasympathetic takes care of itself; it is self-regulating. But stress is metaphorically like leaning on a wall until steady pressure makes it buckle.

As applied to a healing lifestyle, the "airport solution" needs to be activated every day, as follows:

Detach yourself from the stressor. Make sure you have periods of downtime and alone time.

Become centered. Practice meditation, the most desirable tactic, or at the very least find time throughout the day to shut your eyes in a quiet place and take some deep breaths until you feel relaxed and centered. The best breathing technique, which we mentioned in connection with the workplace (page 41), is to breathe in to a count of 4, then breathe out to a count of 6.

Remain active. Getting up and moving around throughout the day stimulates the vagus nerve, one of the principal pathways of the autonomic nervous system. Yoga is even more stimulating and is the best activity for switching from sympathetic overdrive to heightened parasympathetic activity.

Seek positive outlets. In this case, the word *positive* means anything that makes you happy. Making time to be happy is a whole-system strategy, but that's dry and abstract. Happiness is the philosopher's stone for turning a stressful situation into a healing one. In psychological terms, this is why the best way to build a happy life is to build happy days.

Rely on emotional support. Modern society is more and more isolating, which was true even before the Internet and video games greatly accelerated the problem. There is no substitute for emotional bonding, and one thing almost always found in happiness studies is that the happiest people spend an hour or even more per day being in contact, either personally or on the phone, with friends and family who mean the most to them.

Escape if you must. This is generally the hardest choice for most people, who will endure stressful situations long after it is evident that escaping and walking away is the right choice. Aggravated situations like domestic abuse are actually acute stressors. Significant life changes like divorce or switching careers must take many factors into account. However, on an everyday basis you should give yourself the freedom to walk away from heated arguments, malicious gossip, rude e-mails, perpetual complainers, worrywarts, and anyone who is openly criticizing you.

In the end, getting out of sympathetic overdrive, which doesn't occur to enough people, turns out to be the most important single decision you can make, because the benefits to the whole system are lifelong.

6

The Biggest Single Thing to Heal

We've shown you enough proof of the whole-system approach to make a bold statement: mind and body are one. If only one thing is healed in your life, it should be the separation between mind and body. Right now, as most of us live our lives, the self we call "me" hasn't fully mastered the role of the healing self. The chief reason is a loss of wholeness. We have been taught to regard the body as separate from the mind, which is really only a belief. When you look in the mirror, what do you see? Without a second's thought, anyone would say, "My face." But, in fact, your reflection is something you aren't simply looking at—you are reading it.

You read indications of what mood you are in, whether you're feeling fresh or tired, how old you are, and what the years have imprinted upon you. We talked before about an invisible map that we all carry around in our minds about how life and relationships work. But as a visible map your face—and your whole body—symbolize the same things. As your story changes, so does the map. To quote a clever medical axiom, if you want to see what your thoughts were like yesterday, look at your body today. If you want to see what your body will be like tomorrow, look at your thoughts today.

Holistic has become a staple term in the wellness movement, but at a certain level it's easier to live in separation. You are able to detach yourself from what your body is doing. As a sad example, when presented with the possibility that a disease might have a mental component, some patients wail, "You mean I did this to myself?" Self-healing for them comes with an accusation that they are at fault. Yet other people have taken the bodymind to extraordinary heights, and in each case a new possibility is opened that applies to everyone.

EXTRAORDINARY POSSIBILITIES

What if you could wake up every morning for years at a time at precisely the same minute on the clock? This was the case with the pioneering American psychologist William James (although it seems to have been an unconscious ability).

What if you could instantly make an allergy go away? Such a feat has been recorded with patients suffering from multiple personality disorder, where one personality has an allergy that vanishes when a different personality appears. In one case a child would break out in hives while drinking orange juice as his allergic personality emerged at that moment, but he showed no symptoms if another personality was present.

What if you could sit in a freezing ice cave overnight wearing only a thin silk robe? Such a feat has been observed among Tibetan Buddhist monks who have mastered a meditation known as tumo, in which body temperature, which is normally involuntary, comes under conscious control. At one extreme a Westerner who has been submitted to medical testing, Dutchman Wim Hof, performs feats of body temperature control like hiking to a mountaintop in a blizzard wearing only a pair of summer shorts or sitting submerged up to his neck in ice-filled water for several hours.

Hof has his own explanation of how he arrived at his accomplishments: "I said the autonomic nervous system will no longer be autonomic." The problem with this statement is that standard medical understanding holds that the autonomic nervous system cannot be affected voluntarily. A significant study from Holland, however, strongly challenges that understanding and comes down on the side of Hof.

The study, published in 2014 in the *Proceedings of the National Academy of Sciences,* provides evidence for voluntary activation of one activity linked to the autonomic nervous system, the immune response. Healthy volunteers "were trained for 10 days in meditation (third eye meditation), breathing techniques (cyclic hyperventilation followed by breath retention), and exposure to cold (immersions in ice cold water)." The control group was not trained. Then both groups were injected with toxins from a strain of the bacteria *Escherichia coli*, commonly called *E. coli*. *E. coli* normally resides in the intestinal tract and is harmless, but there are pathogenic strains that cause food poisoning, for example.

After receiving the toxin, the trained group followed their voluntary techniques while the control group did nothing. Blood samples were taken, and they revealed that the trained group had a lower release of pro-inflammatory chemicals that the researchers connected to a profound increase in the hormone epinephrine, which is known to decrease inflammation. Besides showing that Wim Hof had grasped the link between his extraordinary physical control and the autonomic nervous system, the Dutch results could theoretically be used by sufferers of persistent inflammation, especially those with autoimmune diseases.

As exotic as these examples may seem, almost anyone can successfully will a red spot to appear on the back of their hands or make their palms grow warm using simple biofeedback. In the age of wearables, medical inventors are seeking ways that we can monitor signs of potential disease or stress through a wrist-worn device that then allows us to return to a normal state of balance at will, again using simple biofeedback.

The big question is what a life built around the bodymind would be like and whether it will create a quantum leap in well-being. We think it will, as it already has for the remarkable woman in the following account.

Tao's Story: Peace and Passion

One couldn't hope for a better example of someone who has lived a healing life than Tao. She has a striking physical presence with her almond skin and dark hair, being the child of a French father and an Indian mother. But it's her personal presence that's even more striking, a kind of smiling serenity all of us would wish for when we reach Tao's age, which is ninety-eight. She is also the world's oldest professional yoga teacher, as acknowledged in *The Guinness Book of World Records*. She stills teaches six to eight yoga classes a week in the New York City area.

When asked what her definition of yoga is, Tao replies immediately, "Union, Oneness."

When asked if she ever plans to retire, she laughingly replies, "I will teach yoga as long as I can breathe."

If you amass the facts of her life, it seems too unique and extraordinary to duplicate. Born Tao Porchon in 1918, she came from a prosperous family in Pondicherry, the French-colonized area of India on the southwest coast. Her mother died giving birth to Tao, and she was raised by her aunt and uncle. When she was eight, she walked into a room of the house to see a man sitting on the floor while visitors and family members touched his feet, the traditional sign of veneration in India.

Tao speaks about all her experiences in a clear, articulate voice. "I was shooed from the room, and then one night my uncle woke me up early. 'We're going on a trip,' he whispered. 'Don't tell your aunt. She'll be worried.' I had no idea where my uncle was taking me, but as it turned out, I joined the first of two marches I made with Mahatma Gandhi. He was the man whose feet they were all touching."

This singular event put Tao on a path she has followed her entire life, whose theme is peace. The casual sight of young yoga practitioners on the beach gave her an early interest in doing yoga herself, even though at the time, the 1920s, the practice was considered to be almost strictly for males. Tao grew up in a spiritual atmosphere, and under the guidance of spiritual luminaries like Sri Aurobindo, who was the most famous guru in that part of India at the time and whose reputation spread worldwide, she developed her own conscious philosophy of life. It is essentially heart centered, seeing love as the universal force that can heal all forms of separation.

Tao believes strongly in going inward and listening to your heart. But there's another path she's walked that seems to contrast with a life of meditation and yoga. Following one central belief—"There is nothing you cannot do"—Tao transformed her inner life into an incredible series of outward accomplishments. To list them all almost strains credibility. She was a peace activist marching with Dr. Martin Luther King Jr. Stage acting, including a career in Hollywood, took up a portion of her life starting in the 1950s, preceded by cabaret singing in London during World War II. But there were other career arcs, as a wife when she became Tao Porchon-Lynch (she was widowed in 1982 and has no children); as a ballroom dancer specializing in the tango, for which she has won hundreds of first-place finishes; and perhaps most unlikely, as a wine connoisseur and writer.

The one thing she shows little interest in is her longevity. "Don't focus on age," she says with a touch of impatience. "It doesn't exist."

There's more, much more, but we're not introducing Tao simply as a remarkable achiever, not even as a beautiful example of a life lived with both peace and passion. Her achievements aren't likely to be duplicated by anyone else, and her historical time, often spent in close proximity to movie stars, writers, activists, and political leaders, won't come again. It's not what makes Tao unique that fascinates us, but what makes her an example everyone can follow. Strip away the uniqueness, and Tao has lived almost a century *consciously shaping her own life.* As a result, if you took a snapshot of any day in her long existence on Earth, you'd see someone who

Put her inner life first

Trusted her feelings and intuition

Valued the now as the source of constant renewal

Cultivated emotional resilience, refusing to be stuck in old wounds and setbacks

Activated her core beliefs, turning her vision into action

Placed her trust in love and spiritual growth every day

We'd call this the model of a healing lifestyle. It's not that Tao hasn't had her share of painful experiences, beginning with the death of her mother; the loss of her husband; and on the physical level, three hip replacements. But instead of converting these experiences into suffering, she has consciously done the opposite—she has become even more dynamic and resilient. One might say that for Tao, only two kinds of experience exist, not the good ones and the bad ones, the moments of pleasure and the moments of pain, but experiences she can celebrate and those she can heal. You can live your life the same way.

Food for Thought

Here's an everyday example of how the separation of mind and body creates practical difficulties in approaching a very familiar problem: weight control. Millions of people flirt with fad diets, and millions more have struggled for years to lose weight and keep it off. How many times have you heard statements like the following? Perhaps you've made them yourself.

"I look in the mirror, and I hate what I see."

"My downfall is chocolate—it goes right to my thighs."

"After my divorce, I gained ten pounds so fast I couldn't believe it."

"I've tried everything, but the weight just won't come off."

"I lose a few pounds, but then I plateau and nothing happens."

These are expressions of a society in which obesity has become epidemic, and in which diets don't work—fewer than 2 percent of dieters successfully lose at least five pounds and keep it off for two years. Magazine editors know they can increase sales by putting a new fad diet on the cover, promising how easy weight loss will be this time, thus feeding a fantasy that the public is all too willing to indulge in. Yet amid all the worry, frustration, self-defeat, and wishful thinking that surrounds weight loss, why would uniting body and mind into the bodymind make a difference?

Because the underlying problem isn't weight. Looking back at the typical statements listed above, all have in common that "I" am unhappy with "it," the body, which turns the normal act of eating into a struggle between what the mind is trying to achieve and what the body is actually doing. Here's what the mind of a dieter is typically doing:

Fantasizing about how much weight the dieter will lose

Believing that just one more try will do the trick

Hating how the dieter's body looks

Struggling to have more willpower

Envying those who have "perfect" bodies

Feeling guilty and ashamed about being overweight

Promising to do better tomorrow

Feeling trapped in bad eating habits that refuse to change

That's an enormous amount of mental activity to dedicate to a project—losing five pounds and keeping it off—that is predictably doomed to failure in the long run. Every bit of this activity is futile because it ignores or weakens the mind-body connection. The mind is disconnected from what the body is actually doing, which looks something like this:

Processing more calories than it needs

Coping with overloads of fat and sugar

Adjusting to toxins in food, air, and water that often have unknown effects

Dealing with low-level inflammation exacerbated by fast food and junk food

Confronting the ups and downs of irregular eating during the day

The separation of mind and body isn't rare or harmless. It lies at the core of why diets don't work. This is a totally unnecessary situation. As a bodymind, each of us is naturally equipped to do what the body and mind both actually want to do: eat normally by following the signals of hunger and satiation.

Two hormones, leptin and ghrelin, are secreted in a natural biorhythm. When the stomach is empty, its cells secrete ghrelin, sending a message to the brain that you register as feeling hungry. When you've had enough to eat, that's the result of a message from leptin, secreted by fat cells, which balances the hunger-satiation rhythm.

In fact, obesity and leptin have both been implicated in risk for Alzheimer's disease. Epidemiological (i.e., population) studies have shown higher circulating leptin levels to be associated with lower risk of Alzheimer's, while lower circulating levels of leptin have been found in patients already suffering from the disease. Leptin receptors are highly expressed in the hippocampus, the area of the brain responsible for short-term memory, which is ravaged by Alzheimer's. Leptin supplementation actually led to less Alzheimer's pathology in this brain region in mouse studies of the disease. This is yet further evidence strengthening the link between the gut and the brain.

Weight Gain: Your Body Isn't to Blame

One reason that early versions of weight-loss surgery were unsuccessful is that when a gastric bypass or lap band procedure was performed, the stomach was left intact, with a small portion sealed off in order to drastically reduce how much food a person could eat at any given time. Instead of consuming a complete meal of cheeseburger with fries, less than a third of that would be satisfying—or should have been. Patients reported being extremely hungry even when their smaller stomach pouches were full, and the reason was that the whole stomach was still secreting a full amount of leptin and ghrelin.

The lesson here is that you haven't had enough to eat when your stomach is full. You've had enough to eat when your brain says so, specifically the region known as the hypothalamus. When mind and body get separated, however, you can override your brain, or rather, you can distort your relationship with your brain. Instead of listening to the natural biorhythm that governs hunger and satiation, you impose your own behaviors. Because you have free will, these can be practically anything, but social norms today are already distorted, which means that you find young children setting in place lifelong habits to which the brain adapts:

Continuing to eat after you're full

Eating excessive sugar and fat

Ingesting alcohol

Snacking at all hours

"Eating your feelings"

Ignoring regular mealtimes

Consuming an imbalanced or very limited diet (such as a diet very low in vegetables and fiber)

Eating too much because you've lost the dieting battle and don't care anymore

Ironically, the blame for the problem is usually placed on the body. The more it gains weight and loses its shape, the more likely a person will judge against his (or more likely her) body for not cooperating. But this lack of cooperation began elsewhere, in the weakened mind-body connection. Let's look a bit deeper. As we already noted, two hormones, leptin and ghrelin, control hunger and satiation. After leptin was first discovered in animals in 1994 by the molecular geneticist Jeffrey Friedman and his colleagues, the *New York Times* led their story with an air of excitement: "It seemed almost too good to be true—a hormone that makes animals, and maybe people, eat less and exercise more. But researchers say that is exactly what they have found." Drug companies were eager from the start to produce a drug to increase leptin levels, which in turn would signal the brain to suppress appetite and increase physical activity.

But the story quickly grew more complicated. To begin with, obese people already produce higher levels of leptin because they have more fat cells than people of normal weight. Why doesn't the leptin curb their appetite? No one knows for certain. The relevant factors include leptin resistance, where the receptors for this particular hormone get overloaded—this is similar to the way overproduction of insulin makes a person less sensitive to it. Leptin and ghrelin are also brain chemicals known as neuropeptides. The brain's receptor sites are probably affected by chronic overeating, but the story gets complicated here because the same region of the brain that contains these receptors, the hypothalamus, is also required to balance overall metabolism and control how much fuel is allocated throughout the body.

Other clues point to problems with the pathways that lead from the hypothalamus after it receives leptin, along with the possibility that not enough leptin gets past the blood-brain barrier. Add to this the genetic factor. Mouse studies dating to 2004 at Columbia University indicated that leptin levels early in life can alter the brain's circuitry, influencing how much food is consumed as adults. This seems to correlate with findings that infants who are fed too much run a higher risk of future obesity. To an extent not yet known, leptin may be able to train the brain by changing its circuitry, leading to altered appetite. A person can wind up at either extreme, being hungry for either too much or too little food.

It's intriguing that leptin can do this, but you are much more powerful in your ability to train your brain, because you can do it consciously. If you want to close the mind-body gap in relation to how you eat, try the following simple mindfulness experiment.

A Practice for Mindful Eating

When you do anything consciously, including eating, you override the brain's default setting and communicate directly with the higher brain, which is responsible for conscious thoughts and actions. Very often we eat unconsciously, without thinking or weighing the consequences of what we're doing.

You can change the situation with a simple mindfulness practice.

The next time you eat anything, whether as a meal or a snack, do the following:

Step 1: Pause before you eat the first bite and take a deep breath.

Step 2: Ask yourself, "Why am I eating this?"

Step 3: Whatever answer you get, take note of it. Better yet, write it down—you might even start a mindful eating journal.

Step 4: Make a conscious choice to eat or not eat.

There is nothing more to do, but this simple practice can lead to major benefits. Your goal is to return to a normal biorhythm of hunger and satiation. When you pause to make a choice, your reason for eating should therefore be "I'm hungry." But there are a host of other reasons we reach for food, like the following:

"I'm bored."

"I can't resist."

"I need comforting."

"There's no use letting all this food go to waste."

"I'm stressed out."

"I feel a craving."

"I'm depressed."

"I'm anxious."

"I don't know why."

"I'm lonely."

"I'm sick of dieting."

"The other people I'm with are eating."

"There's not much left. I might as well finish the package."

"I feel like celebrating."

When you ask yourself why you are eating, it's likely that some of these reasons will come into play. Don't judge against them, and don't force yourself to reject the food out of guilt. Mindfulness is a conscious state, nothing more or less. In this state you are self-aware, and that's the key. When you are self-aware, change comes with less effort than in any other state. The end of unconscious eating is often enough to turn around a person's weight problems, especially if they are mild to moderate.

As you can see, there is hope beyond dieting, a way forward for people who moan "I've tried everything. Nothing works." A whole-system approach to weight loss ends the struggle; no longer is your body the enemy and you its victim.

The Conscious Dieter

We aren't discounting the value of research into how digestion works. An entire research career could be devoted to studying a single hormone like leptin—some already are—and even then, its promise for weight reduction may remain out of reach. (The drug industry offers a host of over-the-counter and prescription diet pills, from so-called fat burners to appetite suppressants, but they are either unproven, ineffective, laden with side effects, or clinically insignificant.)

At the same time, a host of self-defeating beliefs will stop having power over you. In the psychology of overeating, there's a vicious circle at work. Painful beliefs turn into excuses. As an example, take the belief "I've always struggled with my weight, so I must have been born to be this way." No attitude is more self-defeating, and there is science to make this belief stick even harder. There are indeed genetic indicators to suggest that some people are likely to gain weight.

For example, it's invaluable to know the specific gene that modulates how much leptin is produced, because in some obese individuals, a mutation in this gene causes leptin deficiency, leading to uncontrollable weight gain. The story of leptin is interwoven with the story of obesity in the general public, and every clue is worth pursuing. But we covered how misleading the early promise of leptin turned out to be, and in the same vein, the search for a single "fat gene" has been equally fruitless. At best, your genes account for only a percentage of the causes for being overweight—other factors are at work that you can change, such as your psychology, eating habits, and attitudes passed on by your family as a child. These factors are subject to free will when you learn how to change them. The good news is that when change occurs, your genes will respond, altering their activity in the direction of healing this problem.

Many other beliefs keep the vicious circle going by turning guilt into an excuse. How many of the following have you bought into?

The belief that food can make you happy when you're sad

The belief that being full is a state of fulfillment

The belief that known risk factors (high fat, sugar, and salt) don't apply to you—you are protected by magical thinking

The belief, this time unconscious, that the food you don't remember eating doesn't count

The belief that you don't really care how much you weigh

The belief that you don't mind what other people think

These beliefs all deliver a double whammy. They give you an excuse to fall back on, but the excuses stoke defeat. The better the excuse, the worse the failure. In our approach, being realistic means that you break the vicious circle by exchanging fantasy for reality. We are counting on the positive effects of realism, even though many people look at their weight issues and fear the pain of confronting themselves—looking in the mirror is already painful enough.

It takes time, but consciousness is its own reward. In our own experience with meditation, for example, we've seen people return to normal weight effortlessly. The enjoyment of being alive and awake took the place of eating, and when that happens, the whole system begins to normalize. You only realize the futility of fighting your body when you free yourself from duality. You are here to enjoy your thoughts, feelings, desires, and hopes. We've singled out weight issues because they strike close to home for millions of people, and because making weight loss a pleasurable part of personal growth is the last thing most dieters think is possible. But in the larger scheme, conscious living is the goal. Now that we've established how important it is to heal the separation of mind and body, let's unfold the enormous fulfillment that comes from living consciously every day.

7

Mindful or Mindless?

*When we last met Harvard professor Ellen Langer, she was astonish-*ing the psychology world by sending seventy-year-old men into a time capsule and bringing them out seemingly younger. But time travel isn't applicable to everyday life. Having proved her point in spectacular fashion, Langer took up a larger cause: mindfulness. We've been using this term also, showing how being mindful reaches beyond the word's old association with Eastern spiritual practices. Langer has totally Westernized mindfulness with the following definition: Mindfulness, she said before a medical school audience, is the process of actively noticing new things, relinquishing preconceived mindsets, and then acting on the new observations. Our goal here, of unfolding a healing lifestyle, includes the same things.

Langer was very blunt—everyday behavior is mindless most of the time. One of her favorite examples, she said, comes from personal experience: "I once went to make a purchase and I gave [the cashier] my credit card, and she saw it wasn't signed." Langer dutifully signed the card, and the cashier ran it through the machine. She asked Langer to sign the receipt. "[The cashier] then compared the two signatures to make sure they were the same person," Langer recalled. She paused, and it took a moment before the audience caught on and started to laugh. Why would two signatures need to be compared when you've just witnessed the same person signing both? Small instances of mindless behavior tie us to the past and block the possibility of being alive in the moment, alert to possibilities we will never see. In fact, Langer calls her pursuit of mindfulness "the psychology of possibility."

In this book we leap ahead to the ultimate possibility, being conscious all the time. But is that feasible when someone is inundated with deadlines, bills, their children's schooling, and so forth? With so much stress and tension being lobbed at us from so many directions, attention gets dulled. We become reactive rather than attentive. This is how mindfulness gets lost despite our best intentions. Pause for a moment, though, and reflect on how your day has been going. If you're like most people, you will find yourself spending much of your time in reactions rather than paying attention and being mindful of what's going on. As a result, you are living unconsciously while accepting that this is normal. Which of the following things apply to you?

What an Unconscious Day Is Like

You eat irregularly or on the run.

You consume packaged or fast food.

You feel bad about your body or your weight, just as you did the day before and will the day after.

You act rushed and pressured.

You listen to your spouse and kids without really listening.

You react negatively toward someone without examining why this is necessary or right.

You don't notice anything beautiful for the entire day.

You worry about a nagging problem with no plan to solve it.

You habitually take a dim view of the future.

You are haunted by something painful from your past.

You feel stuck and unfulfilled.

You feel insecure or unsafe.

You are lonely.

You lash out at a friend without thinking.

You play the victim.

You fail to stand up for yourself.

You behave like a people pleaser—you go along to get along.

It's amazing how many unconscious reactions and behaviors are taken to be normal. What's really going on is that mindlessness is being normalized. Only by seeing this can you begin to change. There's a major shift toward healing when you decide to stop allowing unconscious negative events to have power over you. There's a spectrum of possibilities here, which can be pictured with a simple diagram:

Unconscious ◄────► Conscious

This is a nonjudgmental way of depicting disaster and fulfillment. At the far left would be a totally unconscious lifestyle, where everything that needs to be healed is unattended to, which ultimately leads to a disaster. At the other end is a completely conscious lifestyle, where every potential problem is paid attention to, making room for complete fulfillment. Few of us live at either extreme, completely in heaven or hell. We find ourselves somewhere in the middle: Sometimes we act unconsciously. Sometimes we act with self-awareness. This gray zone becomes normal and acceptable, without the realization of the damage being caused over time.

Brenda's Story: A Tale from the Gray Zone

Several years ago a woman we'll call Brenda caught a winter cold. The cold didn't go away immediately but kept coming back. She put up with it, annoyed by the hacking cough that developed. Then with great suddenness a fever set in. This, too, she ignored, until one night, the roof caved in.

"I was sitting up in bed, sweating and very weak. My husband is a very caring man. He held me, reassuring me that everything would be okay. But I knew I was very sick. A friend dropped by with some chicken soup. She happens to be a nurse, and after taking one look, she ordered me to go to the emergency room immediately," Brenda recalls.

This intervention probably saved her life, because the ER doctor told Brenda that she had pneumonia in both lungs, which severely compromised her respiration. In fact, she was on the point of respiratory failure and had to be put on a ventilator. The normal treatment with antibiotics will generally knock the infection out, but a blood test revealed that Brenda, unknown to herself, was diabetic. She had struggled with her weight since early adolescence, and type 2 diabetes is very common in obesity. To keep her from fighting the ventilator, the doctors kept her in an unconscious state, sometimes called an induced coma, with massive doses of Valium. It was a drastic course, but they needed to carefully control and monitor her treatment.

"I was stunned. Two days before it was just a cold, and now they were telling me I could die," she says. "My life became a nightmare overnight."

For the next nineteen days it was touch-and-go. Being totally knocked out was interrupted by intervals in which she was brought back into a blurry state of wakefulness. The doctors did this in order to check on how she was faring medically, but Brenda found these episodes terrifying.

"I'd wake up feeling totally anxious, wondering if I was about to die. I had no control over my body anymore, and I was laid out in bed with needles and IV tubes and beeping monitors. It was the worst thing that ever happened to me."

Brenda wasn't prepared for how hard everything was, and when the doctors told her she was out of danger, that the pneumonia was gone, she went home still feeling anxious. She kept telling friends about her brush with death, reinforcing her sense of inner dread and loss of control. In some ways, she'd stayed in crisis mode, which wasn't at all like her.

At fifty-three, Brenda considered herself a strong woman who had spent a lifetime striving to get ahead. She was much more than a survivor. Born in a poor working-class family, she never got past high school, but she knew inside that she was different from the rest of her family. She was more aware of what life could offer, and with determination she set out at eighteen to write her own story.

"I saw girls my age getting pregnant too early, settling into marriages they didn't want, or just thought they wanted. The men had dead-end jobs, most of them, and spent a lot of time drinking beer in front of the TV. It wasn't hard for me to leave it all behind," she says.

She had gone out into the world and made something of herself. In her own mind, Brenda had great self-control. She was a helper, aiding anyone in need. She started a service to cook food for the homeless and was a leader in community support groups. Over the decades between leaving home and reaching middle age, the conscious side of her life wasn't lacking. But after the bout with pneumonia, everything seemed to unravel. Brenda felt depressed and stopped seeing her friends as often. The dinner parties she loved to cook for became much less frequent. On the medical front, her blood sugar came under control with daily insulin injections, but the doctors told her that the damage already caused by her diabetes in some cases couldn't be reversed.

"I was going to three specialists a week. My eyesight was being affected by damage to the retina. I began to have terrible intestinal pain, and they told me it was diverticulosis. My cold feet were due to decreased blood flow to my extremities." Brenda laughs harshly. "I was falling apart. I couldn't believe it."

This apparently sudden drop in the quality of life wasn't sudden—everything had a history. The longest history belonged to the obesity. From it followed a history of elevated blood sugar levels, and other histories for Brenda's damaged circulation, digestion, and eyesight. She deserves sympathy and care—Brenda is seeking both and has gotten them—because the unconscious side of her life is demanding payment. Distressing as her crisis has been, Brenda was leading a normal life with familiar ups and downs. Yet if you look at it realistically, she was deep in the gray zone.

There's an old proverb, "For want of a nail." It comes from a nursery rhyme that was well-known to children in the past, before automobiles replaced the horse:

> *For want of a nail the shoe was lost.*
> *For want of a shoe the horse was lost.*
> *For want of a horse the rider was lost.*
> *For want of a rider the message was lost.*
> *For want of a message the battle was lost.*
> *For want of a battle the kingdom was lost.*
> *And all for the want of a horseshoe nail.*

So what was the nail that Brenda, and many, many other people, lost? They lost the connection to their bodies, to nature, and to themselves.

None of these can be seen in isolation. Your body is how you relate to nature, and when that connection frays, you stop being yourself. By becoming more and more aware that the thoughts you think today and the actions that follow from those ideas and emotions are impacting your body in real time, your life gets transformed at every level. The word *level* is misleading, however, because the bodymind merges everything into a single consciousness of who you are and what is happening to you. Many of us have heard that what's in the past is in the past. Yet your present state is a result of your past. In many ways, the two are inseparable. We can't change the past, but we can change the present.

A skeptic would raise a hand and object. Brenda's story is about medical issues that she had no power to heal. Consciousness is all well and good as a lofty aim, but everyone has medical issues that require treatment. How could someone in Brenda's predicament help herself by the time she was a patient?

This is a polite rendition of a skeptical argument that's often rude and hard nosed. The myth that only drugs and surgery are "real" medicine dies hard. We all wind up going to the doctor's office for various reasons. That doesn't negate the healing self but only gives it another area to deal with.

Worry and the Immune System

As we mentioned earlier, mainstream medicine is biological, believing that the mind must be centered only in the brain. In fact, not just doctors but most scientists who consider the issue would declare that brain = mind. We've been steadily breaking down the validity of this assumption—and it is an assumption, not a fact—by pointing to the intelligence displayed by the bodymind. The information superhighway of the bodymind delivers messages to and from every cell using the same chemicals as nerve cells.

What this implies is that cells are more mindful than we are. Take the immune system as an example. Unless it breaks down in some way, the immune system is never mindless. But if you engage in mindless behavior, the effect is far reaching and can compromise the intelligence of immune cells. Let's go deeper into this interaction by introducing the growing field of psychoneuroimmunology (PNI), which studies the interaction of mental activity with the immune system. PNI is one of the few areas in which the traditional division between physical disorders and psychological disorders isn't compartmentalized. We've already touched on how long-term bereavement can take its toll on a person's health (page 37), including their ability to fend off disease. (During the first two grief-stricken semesters at school after his father's sudden premature death from a heart attack, Rudy remembers being constantly assailed with colds he couldn't shake off. He didn't know at the time about the mind-body connection that turned sorrow into compromised immunity.) The damaging effects of grief can also be quite rapid. In a study involving 100,000 recently widowed spouses, death rates doubled in the first week of mourning.

Since the mind is everywhere, imbuing every cell, PNI could potentially apply to any mental state as it affects the immune system. We'll confine ourselves to a mental activity everyone is familiar with: worry. Worry is mindless in a host of ways. It obsessively occupies the mind and keeps the person in a state of stressful anxiety. It blocks rational solutions to problems and on its own does nothing to provide a solution. Despite its lack of utility, worry is endemic in society. After the 2016 presidential election, for example, Gallup polls showed an immediate sharp upturn in worry. But turbulent politics and presidential races in general increase the public's level of worry. There's anxiety about the future, which is the hallmark of all worrying.

Worry can also serve a purpose. It motivates us to prepare for the worst and to get ready to face an oncoming challenge or threat. This is presumably why worry has been preserved as an evolutionary trait. But when it becomes constant and uncontrollable, chronic worry can be very detrimental to health. Specifically, PNI research has shown that excessive worrying can compromise the immune system and contribute to a host of disorders from heart disease to Alzheimer's. Worry is based in "un-ease" and is just a stone's throw away from "dis-ease."

For anyone who maintains that all you need to do is replace worry with hope, it's important to note that worrying whether something bad will happen isn't the exact opposite of hoping something good will happen. In both cases there is an underlying uncertainty and insecurity about the future, which is accompanied by anxiety. Yet insofar as hope is a positive emotion linked to other positive emotions like optimism and acceptance, the balance tilts greatly in its favor. While chronic negative emotions can literally kill, a positive outlook can help the outcome of diseases as diverse as cancer and HIV-AIDS. Conditions like asthma and psoriasis/eczema can be improved by positive feelings and worsened by stress, depression, and anxiety.

By what mechanisms do these effects, positive and negative, actually work? This is the main thrust of PNI, which traces its origins to the work of psychologist Robert Ader, former director of the University of Rochester's Center for Psychoneuroimmunology Research. In 1974, Ader and his colleagues gave laboratory rats water sweetened with saccharin, followed by a chemical, cytoxan, that suppressed their immune systems and made them nauseous. When the experimenters later force-fed the rats with saccharin, even in very small amounts, the animals subsequently died. The more saccharin they were fed, the more quickly the rats died. Ader concluded that after sufficient conditioning, the taste of saccharin alone was enough to suppress the immune system, directly causing a weakened immune response and thus leading to fatal bacterial and viral infections the rats would have normally been able to fight off. For the first time, the intimate connection between the brain and immune system was being uncovered, giving birth to the field of psychoneuroimmunology, a term Ader coined. The notion that the immune system was totally autonomous had been radically challenged, the start of many findings along the same lines.

In 1981, neuroscientist David Felten, then at Indiana University School of Medicine, later at the University of Rochester, made another major discovery by connecting nerves in the thymus and spleen directly to immune system cells. In 1985, the soon-to-be-celebrated neuroscientist and pharmacologist Candace Pert made a crucial breakthrough by discovering special miniature proteins in the nervous system, called neuropeptides, that interact with both neurons in the brain and immune cells. The effects can be long lasting: neuropeptides strengthen synapses and even change gene expression in both nerve cells and immune cells. Pert's revolutionary studies, which were fundamental to understanding how the body's messaging network operates, went on to prove that neuropeptides were involved with a wide range of activities from social behavior to reproduction, as well as immune response.

Perhaps nothing has been more studied in PNI than the effects of stress, which keeps cropping up in every aspect of how people heal differently. In connection with immunity, chronic stress can suppress the immunity needed for rapidly fighting off infections (innate immunity) or to make antibodies to defend against invading germs (adaptive immunity). Chronic stress has been associated with frequent severe infections and also with worse prognoses for cancer, heart disease, and HIV-AIDS. Worry is a form of self-created stress associated with fear, and in Part Two we devote a lengthy section to stress reduction (page 191).

As intriguing as the findings of PNI have been in pinpointing the biology of the mind-body connection, the power of that connection is still underappreciated. Pointedly relevant here is the case of Norman Cousins, longtime editor of the *Saturday Review*, a peace activist, and in his later years a highly influential author on mind-body healing. His spontaneous discovery that laughter has healing power was once widely discussed, but it's worth repeating the details as recounted on the website Laughter Online University:

In 1964 following a very stressful trip to Russia, [Cousins] was diagnosed with ankylosing spondylitis (a degenerative disease causing the breakdown of collagen), which left him in almost constant pain and motivated his doctor to say he would die within a few months. He disagreed and reasoned that if stress had somehow contributed to his illness (he was not sick before the trip to Russia), then positive emotions should help him feel better.

With his doctors' consent, he checked himself out of the hospital and into a hotel across the street and began taking extremely high doses of vitamin C while exposing himself to a continuous stream of humorous films and similar "laughing matters." He later claimed that 10 minutes of belly rippling laughter would give him two hours of pain-free sleep, when nothing else, not even morphine, could help him. His condition steadily improved and he slowly regained the use of his limbs. Within six months he was back on his feet, and within two years he was able to return to his full-time job at the *Saturday Review*. His story baffled the scientific community and inspired a number of research projects.

Cousins began a personal crusade to publicize his remarkable recovery and its implications for medicine, meeting with popular success but receiving strong resistance from the medical community, which was still a decade away from Ader's experiments with rats dying from the taste of saccharin. Cousins's story is first cousin to the placebo/nocebo effect. Over fifty years later, figuring out how to channel the mind-body connection into healing remains more mystery than science. But perhaps the simplest lesson is the one Cousins learned at the outset, that turning worry and anxiety into laughter can make all the difference.

At the Doctor's:
Be Your Own Advocate

There's a larger lesson to be learned from Cousins's healing, which has to do with not accepting medical treatment passively. As he recounted to a radio interviewer,

> I asked [my doctor] about my chances for full recovery. He leveled with me, admitting that one of the specialists had told him I had one chance in five hundred. The specialist had also stated that he had not personally witnessed a recovery from this comprehensive condition. All this gave me a great deal to think about. Up until that time, I had been more or less disposed to let the doctors worry about my condition. But now I felt a compulsion to get into the act. It seemed clear to me that if I was to be that one in five hundred, I had better be something more than a passive observer.

When the average person goes to the doctor, shows up at the ER, or enters the hospital, the possibility of controlling what happens next is minimal. We put ourselves in the hands of the medical machine, which in reality rests upon individual people—doctors, nurses, physician's assistants, and so on. Human behavior involves lapses and mistakes, and these get magnified in medical care, where misreading a patient's chart or failing to notice a specific symptom can be a matter of life and death. The riskiness of high-tech medicine like gene therapy and toxic cancer treatments is dramatically increased because there is a wider range of mistakes the more complex any treatment is. To be fair, doctors do their utmost to save patients who would have been left to die a generation ago, but they are successful only a percentage of the time.

Risk and mistakes go together, but the general public has limited knowledge of the disturbing facts:

- Medical errors are estimated to cause 440,000 deaths per year in U.S. hospitals alone. It is widely believed that this figure could be grossly inaccurate, because countless mistakes go unreported—death reports offer only the immediate cause, and many doctors band together to protect the reputation of their profession.

- The total direct expense of "adverse events," as medical mistakes are known, is estimated at hundreds of billions of dollars annually.

- Indirect expenses such as lost economic productivity from premature death and unnecessary illness exceed $1 trillion per year.

Statistics barely touch upon the fear involved when any patient thinks about being at the wrong end of a medical mistake. What the patient is all too aware of is the doctor visit that goes by in the blink of an eye. A 2007 analysis of optimal primary-care visits found that they last 16 minutes on average. From 1 to 5 minutes is spent discussing each topic that's raised. This figure is at the high end of estimates, given that according to other studies, the actual face-to-face time spent with a doctor or other health-care provider actually comes down to 7 minutes on average. Doctors place the primary blame on increasing demands for them to fill out medical reports and detailed insurance claims. Patients tend to believe that doctors want to cram in as many paying customers as they can, or simply that the patient as a person doesn't matter very much.

Along these lines, in 2016 renowned cardiologist Dr. John Levinson and Dr. Caleb Gardner published an op-ed piece in the *Wall Street Journal* titled "Turn Off the Computer and Listen to the Patient." They maintained that the introduction of electronic medical records, which are currently mandated by health systems and government, has "degraded relationships between clinicians and our patients" and contributed to the "corporatization—and de-professionalization of US health care." These concerned doctors support the Lown Institute in Boston, a not-for-profit committed to restoring professionalism and a sense of caring to health care. While they believe there is a rightful place for medical computing, all too often doctors have to spend more time dealing with our medical forms than talking with the patient.

As a result, there's a new movement afoot to provide a personal advocate who stays in the doctor's office with the patient. The advocate is basically someone who represents the patient's best interests in any medical situation. The person might be a well-meaning relative who helps an older patient understand what's going on, or who steps in to do attendant tasks like picking up prescriptions and organizing medical bills. But more and more one sees the need for an advocate who is professionally trained to buffer the mounting risks in a health-care system in which less and less time is spent between doctor and patient. If we return to Brenda's situation, the absence of an advocate proved critical in several ways. First and foremost, she wasn't told about the connection between obesity and type 2 diabetes. If that had been done in advance, years ago, the downward spiral in her health could have been prevented. Brenda popped in and out of doctors' offices all the time, but what she was treated for was narrowly defined by the symptoms she exhibited that week. No one put the whole story together. And yet it was not the most complex story, so what happened?

It would be up to an advocate to find out, and needless to say, this has created hostility from some doctors. Used to ruling their domain with absolute authority, few doctors want an overseer in the room asking questions, inserting their own opinions, and potentially finding fault. At worst, the specter of a malpractice suit looms. The movement for professional advocates, which is quite young, insists that looking out for a patient's best interests is benign. The medical profession has its doubts.

The upshot, for now at least, is that patients who want an advocate must play the role themselves. At the heart of the problem is passivity. When we surrender to medical care, whether at the doctor's office, the ER, or the hospital, we shouldn't surrender everything. Poking and prodding is intrusive. Undergoing various tests can be stressful. The minute we walk in the door, we become largely anonymous—a walking set of symptoms replaces the person. There are doctors and nurses who take these negative effects seriously and who go out of their way to offer a personal touch. They should be saluted for their humane compassion in a system that focuses more on impersonal efficiency.

You may like your doctor and feel that he or she cares, but this doesn't rule out being your own advocate. Quite the opposite—the inherent stress in medical treatment is what you want to counter. First comes the stress of worry and anticipation, what is commonly known as white-coat syndrome. We all remember how afraid we became as children thinking about getting a shot from the school nurse or how scary it was sitting in the dentist's chair even before the drill was turned on. Studies have verified that anticipating a stressful situation can cause as great a stress response as actually undergoing the stress. In one study subjects were divided into two groups, one of which gave a public speech while the other was told that they were going to give a speech but actually didn't. Both groups became stressed out, but the researchers wanted to measure how well they recovered from the stress.

Recovery involved three things: having heart rate and breathing return to normal as well as reporting lowered emotional responses like anxiety. The recovery was similar in both groups, indicating that a stress you anticipated but never went through could potentially be just as damaging as the real stressor. Also, how people reported their emotional state as they were recovering was a good predictor of what their heart and respiration were doing. In other words, if you feel emotionally stressed out, your body does, too—no surprise on this point.

How does this apply to going to the doctor's? First, as we mentioned, there is stress in anticipation as well as stress once you get there. Second, under stress people become mentally confused and distracted. Third, the stress is likely to be highest when the doctor or nurse enters the room, at the very moment when you need to be clearheaded. As your own advocate, you don't want to succumb to stress at any point. Your goal is to ask the right questions, get helpful answers, and understand what lies in store for the future. (Everyone knows the frustration of leaving the doctor's office and suddenly remembering all the things you meant to ask but either forgot to or were too stressed out to bring up.)

One big secret to overcoming the stress of medical care is to realize that it is affecting you, and how. Be conscious of the main factors that make any stress worse: repetition, unpredictability, and loss of control. In terms of going to the doctor's or entering the hospital, repetition means that one stressful event follows another, such as undergoing several tests in a row or being asked the same question repeatedly by different people.

Unpredictability means that you don't know what the doctor and the tests are going to reveal. Loss of control means that everything happening to you is dictated by outside forces. Let's face these factors one at a time.

Repetition: In a medical environment you can feel like an object being pushed along a conveyor belt, and at each stop stress gets repeated. It is often unavoidable that a battery of tests must be run or that different people are going to ask you the same question. The worst repetition is probably having to return to the doctor or the hospital repeatedly for the same illness or treatment. One solution is to get off the conveyor belt mentally, which is done by returning to a sense of normal life even though you are in an alien place. Simple measures like chatting with other people, meditating, listening to an audiobook, doing some paperwork, or texting friends—in other words, everyday activity you associate with being in your comfort zone—keep you in the normal world.

Unpredictability: In the age of the Internet, medical care doesn't have to be anywhere near as unpredictable and foreign as it used to be. There's a mountain of information about every aspect of illness and wellness, which millions of people now avail themselves of. The best use of this information is to wait until you know what's wrong with you. The worst use is to anxiously stab in the dark based on the symptoms you have or think you have. When you are at the doctor's or in the hospital, ask somebody to tell you what to expect when it comes to next steps. Waiting passively for an unpredictable event causes stress to mount. (Dentists, who are acutely aware of their patients' anxiety in the chair, now regularly explain in advance the steps of the procedure, offering reassurance along the way. They also try to be realistic about the level of pain or discomfort to come, because sugarcoating this aspect of treatment builds up lack of trust, itself a form of stress.)

Loss of control: Handing yourself over to the ministrations of a stranger is an enormous stress, but it must be done in medical treatments. Knowing that you are going to be in such a situation, there are a number of ways to feel more in control:

- Be informed about your illness. Don't relinquish your opportunity to find out exactly what is wrong with you. This doesn't mean you should challenge your doctor. If you feel the need to inform your doctor about something you saw online, you aren't being confrontational, and most doctors are now used to well-informed patients.

- If the illness isn't temporary and minor, contact someone else who is going through the same diagnosis and treatment as you. This may involve a support group, of which many exist online, or simply talking to another patient in the waiting room or hospital.

- If you are facing a protracted illness, become part of a support group, either locally or online.

- Keep a journal of your health challenge and the progress you are making toward being healed.

- Seek emotional support from a friend or confidant who is empathic and who wants to help (in other words, don't lean upon someone who is merely putting up with you).

- Establish a personal bond with someone who is part of your care—nurses and physician's assistants are typically more accessible and have more time than doctors. Ideally, this bond should be based on something the two of you share—family, children, hobbies, outside interests—not simply your illness.

- Resist the temptation to suffer in silence and to go it alone. Isolation brings a false sense of control. What actually works is to maintain a normal life and social contacts as much as possible.

Following these steps will go a long way to achieving the goal of patient advocacy, which is to serve the patient's best interests at all times. But there remains a difficult unknown, the possibility of a medical error. Studies have shown that adverse events are tied to factors you as the patient can't control, such as fatigue from the long hours and grueling routine that nurses, interns, and residents suffer from. In the rush of hospital routine, it's inevitable that some patients get shortchanged, overlooked, or treated incorrectly.

Writing in the *Journal of the American Medical Association* in 2009, Dr. Tait Shanafelt says, "Numerous global studies involving nearly every medical and surgical specialty indicate that approximately 1 of every 3 physicians is experiencing burnout at any given time." Dr. Dike

Drummond posted on The Happy MD, a website dedicated to alleviating physician burnout, that to understand the problem, think of drawing energy from a kind of bank account in three forms: physical energy (needed simply to keep going); emotional energy (for staying engaged and compassionate); and spiritual energy (necessary for remembering your purpose and why you do what you do as a physician). With every patient there is a withdrawal from this energetic bank account. The trick is to keep enough in the account balance for the sake of the next patient.

As a patient, the more informed and inquisitive you are, the better. If you know what should be happening, you're in a position to spot when something might be amiss. But the whole area of medical mistakes is a fraught one, and the last thing you want to do is to establish an adversarial relationship with the people who are there to care for you. Here's an outline of what to do and what to avoid.

Do

Do be involved in your own care.

Do inform the doctor and nurses that you like to be involved.

Do ask for extra information when you need it.

Do ask for a questionable event, like a pill you aren't sure is the right one, to be checked with the doctor.

Do tell somebody if you have gone out of your comfort zone.

Do remain polite in all of the above.

Do praise the doctor and nurses when it's called for. A show of gratitude doesn't go amiss.

Don't

Don't act hostile, suspicious, or demanding.

Don't challenge the competency of doctors and nurses.

Don't nag or whine, no matter how anxious you are. Reserve these feelings for someone in your family, a friend, or a member of a support group.

Don't pretend you know as much (or more) than the people who are treating you.

Don't, when hospitalized, repeatedly press the call button or run to the nurses' station. Trust their routine. Realize that the main reason patients call a nurse is more out of anxiety than out of real need.

Don't play the part of a victim. Show your caregivers that you are maintaining a normal sense of security, control, and good cheer even under trying circumstances.

Probably the most important finding about medical mistakes is that they are frequently caused by lack of communication. The breakdown mostly occurs between doctors and other medical staff as orders get passed along that are confusing, impatiently or poorly stated, or otherwise lost in translation. As before, if you are well informed about your illness and treatment, you are in the best position to spot miscommunication. But to be realistic, this isn't an area where patients can do very much, and the medical profession has proved very reluctant to deal with the problem, or even to recognize how serious it has become.

Social categories matter here. Medical errors are the most likely to occur under conditions in which some groups are already disadvantaged. If you are elderly, poor, have less education, or belong to a racial minority, you aren't in the same position as someone who is younger, white, highly educated, and well to do. To no one's surprise, privilege has its privileges. But even so, each of us should take responsibility for what a patient can actually do to minimize miscommunication, which means

Being clear about how you describe your symptoms.

Stating your expectations in a realistic way. Do you want relief from pain, a cure, some signs of progress, or reassurance that the worst isn't going to happen to you? Patients have different expectations, and you need to be clear about what yours are so that the doctor and other caregivers know.

Speak up when you don't understand something about your condition.

Ask about side effects of the drugs prescribed to you.

Don't be afraid to inform someone if you didn't get answers to your questions.

We've gone into detail on this topic because it stands in for something bigger than going to the doctor's office. When you become an advocate for yourself, you are being conscious instead of unconscious, you are valuing self-care at least as much as care from someone else, and in this attitude of self-care, you are being a healer.

The patient advocacy movement arouses suspicion among doctors because it overturns the routine of patient care that puts the doctor completely in charge. Change is likely to come slowly, with organized resistance on the other side. Likewise, if you tell a doctor that you are a healer, his reaction isn't likely to be positive. He'd assume, more likely than not, that you are usurping his role. But we hope you believe, after reading this far, that no such challenge is being mounted. Our bodies already make us self-healers. Our conscious and unconscious choices determine whether we are helping or harming the healing response.

The most basic choice of all is to be mindful instead of mindless. Nothing about this decision represents an attack on the medical profession. If you get involved in your own care and a doctor or caregiver acts offended, they are committing a medical mistake already. Good doctors welcome patients who are not only involved but show every sign of both self-reliance and trust in the doctor. The two attitudes aren't mutually exclusive, because in the end, doctors and patients are joined in the same activity, promoting the healing response as best they can.

8

The Hidden Power of Beliefs

In the steady progress of artificial intelligence, computers will never acquire one of the leading features of human intelligence: belief. A computer's reality is entirely based on facts that can be turned into the digital language of ones and zeros. If cold calculation represents a better mode of thinking than our messy mix of reason and emotion, even that is an assumption that AI (artificial intelligence) champions believe in and skeptics don't. Belief makes us more human—no other creature exhibits this quality of mind—but how beliefs actually work is still a mystery.

Which of the following sayings is truer: "I'll believe it when I see it" or "You'll see it when you believe it"? Neither one, because despite the fact that they are opposites, each statement is a half-truth at best. Every physician in practice has experienced the phenomenon of patients who die from their diagnosis. This refers to how hearing bad news can be so traumatic, a patient quickly declines even though the underlying condition may be treatable, or at the very least allow for months and years of survival. Seeing isn't believing in these cases. Two people diagnosed with lung cancer can be shown x-rays that are essentially identical, and yet their survival can't be predicted, nor would an oncologist expect it to be the same for both of them.

There's an old medical school joke about the woman who goes to the doctor for her annual checkup and tells him that she is afraid she has cancer. He puts her through a battery of tests and informs her that it's good news. She's completely healthy with no signs of cancer. She returns the next year and again tells the doctor she thinks she has cancer, and once more the tests reveal no signs of malignancy. This goes on for decades. Finally the time comes, when the woman is seventy-five, that the doctor says, "I'm sorry to tell you this, but you have cancer."

"I told you so!" the woman exclaims in triumph.

How do we manifest our beliefs into a physical condition? That's the key mystery. Many doctors would cling to an entirely physical explanation, pointing to changes in the immune system or the brain. But those changes in our physiology are the evidence of what the bodymind is doing. Once you start dealing with beliefs, the elusive word *why* cannot be avoided. For example, it is well documented that traumatic psychological events like being fired or losing a loved one lower a person's immune response. It has also been shown that the brain by itself can stubbornly "believe" in a falsehood. This is what happens in the phenomenon known as phantom limbs, where an arm or leg has been amputated, but the patient continues to feel the shape of the missing limb. This phantom is often accompanied by pain and discomfort. Even though the mind knows the truth, the brain is holding on to the physical equivalent of a belief.

Quite often when our bodies do something we don't like, "Why?" becomes more personal: "Why is this happening to me?" No answer that depends on purely physical causes is dependable. Even in simple everyday cases, a whole-system approach is needed. In the winter, for example, catching a cold isn't just a matter of being exposed to the cold virus (rhinovirus). Some people have "emotional immunity" that protects them even if you give them a pure dose of rhinovirus directly into their noses.

That's exactly what a team from Carnegie Mellon University and the University of Pittsburgh did with 276 adult subjects. The virus entered the bloodstream and infected almost everyone. But only a certain percentage actually developed cold symptoms. Why? The researchers thought that the difference was created by relationships. This turned out to be right, and not just that, the effect could be quantified. Subjects were asked about how many social relationships they had, ranging from family and friends to clubs, schoolmates, church, and volunteer work—a total of twelve categories. One point was assigned for each relationship in which the subject made contact, either in person or by phone, at least once in two weeks, so the maximum score was twelve.

The key finding was that people who reported only one to three kinds of relationships were four times more likely to exhibit cold symptoms than those who reported six or more kinds of relationships. It wouldn't seem all that surprising if someone with a loving mother who supplied sympathy and chicken soup had greater immunity than a lonely widower. But this study proved a bit baffling because it was the number and diversity of relationships that counted, not their level of intimacy. Being embedded in a wide social network created emotional immunity, even when physical risk factors like smoking, antibodies, exercise, and sleep were accounted for.

As before, finding evidence of a bodymind phenomenon isn't difficult. It's the "why," which originates in the mind, that needs to be unraveled. The best place to begin is with the placebo effect, a term almost everyone is familiar with but one that has yet to be explained, despite years of trying.

Being Your Own Placebo

Medical advances depend on knowing reliably what works and what doesn't. No one wants to take an ineffective drug or supplement. Perhaps you're considering an over-the-counter homeopathic product. Will it work for you? Is it acceptable if it works for only a percentage of people? These are basic questions, but there is an X factor to consider. If you take the homeopathic and feel better from it, maybe the physical product isn't actually causing the improvement, but only your belief that it would.

The fact is that your beliefs, your conditioning since childhood, and even the genes you inherited from your parents belong to the X factor. The homeopathic—or any other drug or supplement—only partially determines how you will respond to treatment. The placebo effect, which brings healing totally without an active medical ingredient, is very enticing. If you could be your own placebo, the safest form of healing would be at your disposal. Every cell in your body knows exactly what it needs and takes nothing else. Could this be true of the bodymind as a whole?

If so, we would only need to contact the level of the self that totally supports our cells, consciously offering what they need.

Before deciding if that's a realistic possibility, let's look deeper into the whole phenomenon. Among physicians, the placebo effect has been fascinating, confusing, and frustrating at the same time. In practice, few if any doctors would dare to treat a patient with a dummy pill in place of the real thing. But in clinical drug trials the placebo effect must be ruled out; otherwise, the efficacy of the drug can't be known.

The word *placebo* means "I shall please" and was once used in a prayer—*Placebo Domino* means "I shall please the Lord." The religious association survives even today, because some of the placebo effect comes from the ritual of receiving a medication at the hands of a white-coated physician in the environment of his office or the hospital. It wasn't until the eighteenth century that the term *placebo* was used to mean a decoy drug. In modern medicine, a pioneering American anesthesiologist, Henry Knowles Beecher, carried out some of the earliest studies of the placebo effect in the 1950s.

Beecher had served on the front as a doctor in World War II, where he observed that some severely wounded soldiers apparently felt so little pain that they asked for no painkillers. Based later at Harvard and Massachusetts General Hospital, Beecher put forward in a seminal 1955 paper that the *perception* of pain isn't always dependent on the severity of an injury or illness. Today, this principle is widely accepted—we know that the only reliable way to measure the level of pain is by asking the patient to rate his own pain. What seems like a 10 to one person on a scale of 1 to 10, where 10 is excruciating pain, will be rated a 7 or even lower by another patient.

Beecher asked whether the perception of pain can be influenced by someone's beliefs and expectations surrounding a placebo drug. He went on to conduct a series of clinical studies to test his hypothesis, eventually concluding that in roughly 35 percent of successful treatments, the placebo effect was at work. This finding rocked the medical world at the time, and the shock waves only spread wider in the coming decades. In later studies, it was found that the placebo effect was even more prevalent, accounting for up to 60 percent of therapeutic effects. For example, in a study of leading antidepressants (chemically known as fluoxetine, sertraline, and paroxetine), 50 percent of the positive outcomes were due to the placebo effect and only 27 percent to the drugs themselves.

Beecher's findings also had indirect effects on the medical profession. For example, over the next half century, doctors' attitudes about telling patients the truth changed radically. It had been accepted practice not to inform a patient of a fatal diagnosis, for fear that the news would do more harm than good. When Emperor Hirohito of Japan was diagnosed with intestinal cancer in September 1987, he wasn't told by the court physician and had no idea of his condition up to his death in January 1989, more than a year later. We live in an era of truth telling today, and every patient expects as a matter of course to be told their diagnosis. But the other side of the placebo effect, the nocebo effect, is also real; the belief that things are going to turn out badly exerts its own powerful influence.

The nocebo effect can be self-induced. In a 2017 study of over 1,300 patients who had been diagnosed with nonceliac gluten sensitivity, 40 percent turned out to have no actual sensitivity to gluten when tested in a blind clinical trial—that is, the patients didn't know whether they were receiving actual gluten or not. In only 16 percent of the cases were gluten-sensitive symptoms found despite previous diagnostic testing. It was concluded that the 40 percent with no symptoms after receiving gluten may have experienced the nocebo effect in everyday life when they showed symptoms.

A strange and fatal form of the nocebo effect occurred in an epidemic of people dying in their sleep in the 1980s. In this case, dozens of Southeast Asian immigrants, men in their early thirties, began dying, always in their sleep, thousands of miles away from their homeland. This mysterious occurrence centered on males of the Hmong people (also known as the Mong), a mountainous tribe spread from Laos to China. Here, they also had in common that they originally came from Laos. This medical mystery acquired the name of sudden unexpected nocturnal death syndrome (SUNDS).

Interviews would later reveal that the Hmong men were being killed by their belief in the spirit world. They were dying in their sleep of heart attacks while they were being literally "scared to death," as survivors reported the experience. The common feature medically was "sleep paralysis," a harmless natural occurrence that happens to everyone in deep sleep, in which the limbs become immobile. But here the paralysis was part of a lucid, or waking, dream, in which the dreamer believes he's no longer asleep and finds to his horror that he's paralyzed.

Sleep paralysis has long been associated with nocturnal evil in many cultures. In Indonesia, it's called *digeunton* ("to press on"), in China *bei gui ya* ("held by a ghost"). The word *nightmare* is derived from the Dutch *nachtmerrie*, in which the "mare" is a female supernatural being who lies on dreamers' chests and suffocates them. In many parts of the West, it's called "Old Hag" syndrome. The terror of being unable to move, combined with the preexisting cultural belief that this state involves an evil being, was enough to induce cardiac arrest among the Hmong men.

Because a placebo is a two-edged sword, various issues arise. For instance, can a placebo wind up having the reverse effect, actually harming the patient? Consider one study that pushed the envelope on the efficacy of antidepressants, suggesting that up to 75 percent of success in treating depression was due to the placebo effect. The public has reacted badly to the headline "Antidepressants Don't Work," because it undermines their faith in a pill they wish worked. Being told that they were in essence duped left many feeling alone, isolated, and helpless in the face of their depression. But drug dependency has remained at high levels for this disorder. Prozac Nation wasn't prepared to turn into Placebo Nation—and it didn't. The market for the most popular antidepressants has only increased.

Moreover, you don't even need a pill to create a placebo effect. When patients with irritable bowel syndrome underwent a sham acupuncture treatment in which the needles never actually penetrated the skin, the researchers found that a striking 44 percent of subjects reported improvement of their symptoms, including the digestive problems and pain associated with irritable bowel. Moreover, when the sham procedure was combined with positive reinforcement and encouragement from the acupuncturist, fully 62 percent of subjects reported improvement.

For a long time, explaining the placebo effect was like entering a black box, a scientist's term for a phenomenon that doesn't connect cause and effect. Here, no one knew what happened between administering the placebo and later observing its effect, or lack of effect. Medicine was hampered by an apples-and-oranges mentality, because the physical nature of a sugar pill (the apple) didn't match the psychological nature of whatever the pill was doing (the orange). With a whole-system approach we aren't saddled with this dilemma, because there is no black box. The placebo effect works because it crosses the artificial boundary between mind and body. Here's a diagram to show what happens:

Placebo ⟶ Interpretation ⟶ Outcome

Nothing truly foreign is occurring in the placebo effect—quite the opposite. *Any* experience you have undergoes interpretation before it has *any* effect. In a classic placebo experiment, patients suffering from chronic nausea were given a drug they were told would make their nausea go away, and in around 30 percent of cases this happened. What they weren't told was that the drug in question actually causes nausea. So the power of interpretation went beyond sugar pills that improved symptoms despite the pills' innocuousness; now experimenters saw improvements despite the physical action of the drug. As shocking as Beecher's original findings were, this new wrinkle has yet to impact the medical profession as much as it should.

There's no need to unfold a litany of the numerous placebo/nocebo trials that together reach the same conclusion. But to underline how indisputable the effect is, we want to show how blanket the effect is—it extends beyond sham drugs to procedures as drastic as sham surgery, and the whole system is potentially subject to the placebo effect, far beyond the initial observations about pain. Here are some research highlights:

- In a 2009 trial, patients were treated to alleviate the pain due to osteoporosis; the procedure consisted of repairing damaged vertebrae with injections of bone cement. The placebo group didn't receive the injection; instead, the doctor put pressure on their spines while allowing them to smell the medical cement. Both groups reported the same level of pain relief. In the end, the placebo results helped to invalidate the efficacy of the actual procedure, since shown to be effective, it would have to outperform a sham procedure. (This still begs the question of how valuable it might be to relieve pain strictly by relying on the placebo effect.) But what if those on placebo also benefited simply based on their belief in the promise of the treatment?

- In his career as a leading Alzheimer's researcher, Rudy has tracked the disease at the genetic level to discover how to combat the senile plaques that litter the brains of Alzheimer's patients and destroy nerve cells. He actively participates in the urgent mission to develop drugs to stop the buildup of these plaques. A small trial was conducted on a promising new medicine developed in Australia called PBT2, which was aimed at clearing existing plaque away.

  As a placebo, subjects were given an inert red pill, and plaque levels were assessed before and after the treatment using brain imaging. Subjects given the PBT2 did show less plaque on average after taking the drug, but so did the placebo group, to a lesser extent. Unfortunately, this was enough to rule the trial drug a failure. In the larger picture, however, it's remarkable that a placebo can lead to an actual physiological change, not simply a patient's subjective sense of reduced pain and discomfort. If belief and expectation can alter the brain, how far does their power extend?

- Because belief is so crucial in the placebo effect, it would seem that deception is necessary in order to trigger a patient's trust in the actual drug he might be taking. But Ted Kaptchuk, a leading placebo researcher at Harvard, is investigating the possibility of doing away with the element of deception. He tells patients up front that they will be taking a placebo, but he also informs them about how powerful and effective placebos can be. In one trial with irritable bowel syndrome, it was found that roughly 59 percent of those on an open placebo treatment reported improvements, as compared with 35 percent of the control group that received no treatment. The results may seem modest, but they make the point that giving a placebo isn't the same as doing nothing, a lingering prejudice that still exists in some medical quarters.

How to Change Your Interpretation

Although the actual cause of the placebo effect is unknown, there is no doubt about the crucial role played by belief, expectation, and perception. As far back as 1949 the pioneering researcher Stewart Wolf proposed that the placebo effect was greatly influenced by personal perception, writing that "the mechanisms of the body are capable of reacting not only to direct physical and chemical stimulation but also to symbolic stimuli, words and events which have somehow acquired special meaning for the individual."

We generally don't think that symbols shape our sense of reality, much less shape our bodies. But when you were sick in bed as a child, by taking a small white pill with the expectation of getting better you entered the world of symbols. The contents of the pill were unknown to you, but it symbolically stood for getting well, and more symbols began to draw a map in your head. Pause for a moment and see if you agree or disagree with the following statements about doctors:

Agree/Disagree: If you want to stay well, keep away from doctors.

Agree/Disagree: Health-care costs will never go down if doctors have anything to do with it.

Agree/Disagree: Doctors ask us to trust them, but that's hard when one medical study says A and the next study says the opposite of A.

Agree/Disagree: Doctors are in the pockets of the big drug companies.

Agree/Disagree: Most doctors want to get you out of their office as fast as possible.

Rationally speaking, each of these statements can be fact-checked, but most people will immediately agree or disagree based on other criteria—their good or bad experience with medical care, stories they get from the media, biases acquired from friends and family, good or bad feelings toward upper-income earners, and so on. We rarely bother to untangle our personal reasons for making snap judgments, and yet we also don't want to back down when facts are actually brought to bear. Those negative statements about doctors turn them into symbols of bad personal traits: greed, incompetence, selfishness, callousness, and even outright dishonesty.

Symbols are so powerful that the rational mind has a difficult time laying them to rest. Compared with the certainty and simplicity of a snap judgment, the counterarguments against those same negative statements are boring and cautious. For example,

Not all doctors fit a blanket description.

It would take statistical studies to verify how many doctors actually do exhibit these bad traits.

Such studies are likely to be unreliable because judgments are so subjective.

For each bad thing a doctor is accused of doing, he deserves to tell his side of the story.

We could have presented a set of positive statements instead that turn doctors into positive symbols of professionalism, education, caring, dedication, compassion, and selflessness. An unscrupulous Medicare cheater committing fraud in a Florida pill mill symbolizes something very different from a member of Doctors Without Borders combating an Ebola outbreak in West Africa. Depending on how you have responded since childhood, these symbols have become ingrained inside you as belief, habit, conditioning, fears, and preconceived notions. All exist on a sliding scale between very positive and very negative. Therefore, placebo/nocebo extend far beyond the usual definition. No one can say with authority how a symbol, seemingly so abstract and impalpable a thing, can create changes in the physiology. But our discussion of the placebo effect leaves no doubt about the fact that personal experiences get metabolized as much as food, air, and water get metabolized. In theory you could follow every molecule in a bit of broccoli to see where it wound up in the body, but that's not true of an experience, because the mind-body links are invisible to begin with and only trigger physical effects at a later stage of chemical reaction.

Alia Crum, a Stanford psychologist who led a 2017 study calling for more research into the nonphysical aspect of healing, encapsulates the issues perfectly: "We have long been mystified by the placebo effect," she says. "But the placebo effect isn't some mysterious response to a sugar pill. It is the robust and measurable effect of three components: the body's natural ability to heal, the patient mindset, and the social context. When we start to see the placebo effect for what it really is, we can stop discounting it as medically superfluous and can work to deliberately harness its underlying components to improve health care."

Being your own placebo obviously would be impossible if you had to fool yourself. Kaptchuk's tactic of eliminating the element of deception opens the door to a different approach, but even here, the symbolic effect of a physician telling you that a placebo can be powerful medicine counts for something. Telling yourself that you might get relief from a migraine headache or lower-back pain by taking a sugar pill isn't viable—you might as well dispense with the pill and drink a glass of water. But you can activate the placebo effect by instilling the one thing that triggers it: a positive belief in being healed.

The Qualities of a Healing Belief

It must be convincing enough to inspire confidence.

It must dispel a negative belief.

It must have personal meaning for you.

It has to bring positive results.

It must be reliable and repeatable in its effects.

Each of these criteria is realistic and nonmystical, although some would argue that the potent effects of phenomena like faith healing, psychic healing, and indigenous cults like voodoo are entangled with everything on the list. Taking no position on faith healing, we endorse the power of having faith in yourself—that's where control of the healing response comes from. But there are many levels of confidence a belief can instill. Hearing a friend say "I'm sure you'll get better" has very little effect compared to having a highly respected physician tell you the same thing. Yet nothing is as powerful as building your own belief system and, more important, knowing that everyone's belief system is dynamic and can change on a dime. Imagine that you are going to a party at the invitation of a friend, and knowing nothing about what to expect, you ask on the way to the party who is going to be there. Your friend might reply:

"Just some boring people from the office."

"The Broadway cast from *Hamilton*."

"A group of civil rights activists."

"Some convicts who just got out of prison and want a fresh start."

These responses will have wildly different effects on your belief that you will enjoy the party or not. They trigger everything we've been talking about with the placebo effect, including expectation, perception, outcome, and symbols. We gather them together under the heading of interpretation, the process that turns the data of an experience into an actual experience. Interpretation can be thought of as a series of filters. When you set eyes on something, hear words, or confront an everyday situation, your filters ask:

Do I want this experience or should I shut it out?

How much of this is going to feel good? How much
is going to feel bad?

Have I been here before? If so, how did I react?

Is this something that needs my attention right now?

Do I need to say or do something?

Do I even care?

Once you are conscious of the process of interpretation, you can change any aspect of it. At the opposite extreme you can react mindlessly with a default response, like a child who hates spinach no matter how much coaxing her parents do. The point in this chapter is that your cells are listening to your interpretation and respond to it *as their own experience*. "I hate spinach" can trigger a gag reflex (or pretending to have one), and by association the parents' blood pressure could go up.

So let's accept that your beliefs, and all the other elements of interpretation, are activating bodily responses all the time. It's your conscious choice to make any situation a healing one, beginning with your choice of beliefs. The most healing beliefs would include:

I expect to be happy and well.

I'm in control.

Whatever needs to be faced, I'm up to the challenge.

I feel safe and unafraid.

I'm supported by family and friends.

I am loved and I love in return.

I accept who I am.

Notice that only the first belief, "I expect to be happy and well," touches upon health and then only by implication. The other beliefs pertain to how you relate to yourself. The main point of this book is that everything comes down to the self. In large part, your "self" *is* a system of beliefs. A belief isn't like a coat you can put on or take off. It's more like a piece of invisible genetic code that has helped to create you as a person.

Among the beliefs that hinder healing, almost all are simply the reverse of a healing belief:

I expect to be unhappy and get sick more often than most people.

I'm not in control of my life—a lot depends on people and circumstances beyond my control.

Many challenges would be too hard to bear.

I'm subject to anxiety and worry about the bad things that could happen to me.

I'm pretty much alone and have to look after myself without support from others.

I don't have much love in my life.

I judge against myself.

We realize that, on both the positive and negative sides of the equation, these statements aren't couched in the form "I believe X." But peel away the sentiment being expressed, and even something that looks like a statement of fact ("I accept who I am" or "I'm subject to anxiety") can be traced to a belief that is being masked. "I'm subject to anxiety," for example, could mask any manner of belief, such as "The world isn't a safe place," "It only makes sense to be afraid," "Fear keeps me vigilant and alert," or "I was made this way," to give just some leading possibilities.

The process of transforming your beliefs in the direction of healing isn't mysterious. It just breaks down negative beliefs step by step.

1. When you catch yourself saying anything that's negative, ask, "Is this really true?" Exposing an automatic reflex to rational questioning is an important step in letting go of it.

2. When you start to examine a negative belief that has come up, ask, "Is this really helping me?"

3. Be detached from the negative beliefs of others—secondhand infection happens frequently.

4. For every negative belief you expose, offer yourself two positive beliefs.

5. Journal about your journey of self-exploration. Write down any changes in your belief system that you see are happening, or simply that you want to happen.

6. Spend more time with supportive, loving, inspiring, and generally positive people. Avoid the opposite kind of people.

7. Value the whole project of self-care and increasing your state of well-being.

We want to particularly reinforce step 4: *For every negative belief you expose, offer yourself two positive beliefs.* This is a powerful way to become the creator of your own belief system. If you don't, you will passively accept all kinds of secondhand unproven beliefs that will continue to be in charge. Try creating new beliefs by writing them down, and take the time to choose beliefs you have some confidence in, not random or abstract possibilities. For example:

Negative belief: I can foresee the worst-case scenario, and it's bound to come true.

New beliefs: I can't really see the future. Dwelling on the worst-case scenario doesn't help me. If I am open to other possibilities, I'm more likely to find a better outcome. A lot of times I've thought the worst-case scenario would play out and then it didn't.

Negative belief: I'm no good in a crisis.

New beliefs: Asking for help doesn't mean I'm weak. I can learn how to deal with this crisis by consulting someone who has survived the same thing. No one says I have to do this alone. I've survived a lot so far. A crisis can also be an opportunity. There's a solution to every problem if you look deep enough.

Even though each of these beliefs, whether negative or not, isn't the same as a statement of fact, it has the magical property of turning into a self-fulfilling prophecy. Reality goes where belief leads. How? For a group of medical researchers the question may come down to genetics, which can strongly influence one's predisposition to the placebo/nocebo effect. Predicting who is more likely to benefit from a placebo is potentially very important for clinical trials of new drugs. In the search for genetic links, a set of genes has already been termed the "placebome," in keeping with the term *genome* and more recent offshoots like the *microbiome*. The identification and characterization of the placebome is still in the earliest stages, but some compelling clues have already come to light. Genes involved with the brain neurochemical dopamine, which is associated with risk taking and reward, as well as other genes involved with opiates, pain relief, and even cannabinoids (molecules produced by the brain that are analogues of the active ingredients in marijuana), have been implicated. Given the whole-system nature of the placebo effect, a complex web of processes is very likely at work, going all the way to the level of our genes.

No one knows where the genetic trail might lead. In the meantime, for anyone leading a healing lifestyle, negative beliefs must be exposed to light before change is possible. It's fascinating to discover that the words we apply to our sense of being conscious, like *alert, vigilant, self-aware,* and *awake,* also apply to our cells. As we'll see in the next chapter, the healer's role is much easier to adopt once you realize that you are just expanding one of nature's greatest gifts, the wisdom of the body.

9

The Wise Healer

It's a major step to realize that body and mind should be considered the bodymind, thus curing the state of separation. But we can go a step further and reach a deeper level of healing. Not only deeper, but easier and more natural. This step involves the wisdom of the body, which too many people ignore or don't believe in. If your body and mind were law partners, they'd hang out a shingle with Mind, Esq., first, because the general consensus says that the mind deserves to be the senior partner.

In contrast, the body supposedly understands nothing. The image in the Book of Genesis of God creating Adam out of a lump of clay has had a far-reaching impact, even though the lumps are now clumps of cells. Pause for a moment to answer the following questions:

Which is smarter, the body or the mind?

Which is more creative?

Which is wiser?

Which are you prouder of right now?

If you think more highly, right this minute, of your mind than your body (anyone past the age of fifty will sympathize), then you have bought into old beliefs that need revising. The intelligence of the body is millions of years older and far deeper than the rational mind. The body deserves to be equal in the bodymind partnership. The practical result of this realization is to become a wise healer.

Control by the Host

The wisdom of the body exists everywhere. Scientists and laypeople alike have developed the habit of exalting the brain as the only place where intelligence resides. By now, having seen how the body's information superhighway works, you know that messaging is a constant process that involves all 50 trillion cells. But this doesn't dispel the worshipful regard that has been directed at the brain. After all, can a liver cell compose Beethoven's Fifth Symphony? Can a kidney cell understand $E = mc^2$? Actually, the body performs feats of intelligence that leave those two examples in the dust.

Since our topic is healing, let's look at the central player in protecting the body from harm, the immune system. In medicine we talk about "control by the host," which means that after a disease organism (pathogen) enters the body, only a certain percentage of people will become infected, and only another percentage will exhibit symptoms. The reason that everyone doesn't get sick is that the body runs a multilayered defense to control everything happening inside us. Control by the host is a natural phenomenon that begins with a series of physical defenses dating back tens of millions of years. An open wound exposed to the air puts you at high risk for invading pathogens, yet if you stop and think, your lungs are just as exposed and open to the air as an open wound.

The difference is that the respiratory system is lined with mucus, which catches invading dust and germs like flypaper. In addition, each breath you take must follow a long, convoluted path before reaching the delicate membranes where oxygen is exchanged with carbon dioxide, and all these byways block or trap more invaders.

The skull and spinal vertebrae are formidable defensive walls, because few pathogens can make it through bone. Your skin is a much softer barrier, but it is more defensive than you might think: the dryness of the skin's surface and the salt left behind from sweating make a poor environment for pathogens to multiply in. Where the skin has natural openings, there are other defensive measures like the flow of tears that drain debris from the eyes and the acidity of vaginal secretions. Tears, saliva, and nasal secretions contain lysozyme, an enzyme that breaks down the cell walls of bacteria.

Inevitably this first line of defense proves inadequate, because the same process of evolution that creates defense mechanisms also creates invaders better equipped to get past them. When a pathogen makes it inside the body, usually as a bacterium or virus, hand-to-hand combat is necessary. You'll remember video footage of white cells moving in on an invader, surrounding it, and gobbling it up. The immune cells responsible for this are macrophages (literally "big eaters"). Behind a process that looks as crude as a boa constrictor swallowing its prey whole lies incredibly complex chemical messaging. We'll dwell on just one aspect in order to show that calling the immune response intelligent isn't an exaggeration.

Take the simplest example: catching a cold. Everyone assumes that catching a cold is a physical process. The person is exposed to the cold virus, it enters the bloodstream, usually by being inhaled, and as the virus multiplies, a struggle goes on between it and the body's immune system. In a healthy child or adult, the immune system wins. The bloodstream is littered for a few days with toxins from the virus, along with the residue of dead viruses and the dead white cells that gobbled them up. In a week the body is cleared of invaders, new antibodies have been formed to protect against the same virus trying to sneak back in again, and you are once again well.

The whole thing looks absolutely physical, but descend to the meeting point where a single cold virus, inhaled on a chilly day by your third grader on the way home from school, comes face-to-face with a single macrophage, the foot soldier of the body's immune system. A fight is about to happen, yet it can't without two packages of knowledge clashing first. One package of knowledge is built into the cold virus's DNA; the other is built into the child's DNA. When they meet, each package has an information exchange with the other. If the cold virus, which is one of the fastest-mutating organisms on the planet, has brought something new to the table, the macrophage is completely baffled. It doesn't know what to do.

For the moment, therefore, the superior knowledge of the cold virus succeeds in doing what it wants to do, which is to trigger more of itself in vast numbers throughout the bloodstream. But the body's heal-ing system is millions of times smarter than a cold virus, and it can adapt to change even faster than a virus can mutate. Back at immunity headquarters—the lymphatic system that has its own pathways separate from the bloodstream—an urgent message has been received. The mac-rophage informs the immune system exactly what the new chemical is that it can't block, usually a protein.

Now a specific type of white cell, known as a B-cell lymphocyte, revs up into "hypermutation," and in short order it produces a single antibody coded to block the sly protein that let the cold virus get past the body's defenses. It has taken decades of medical research to uncover and describe these minuscule processes (which also involve killer T-cells, helper T-cells, and others as well), but the key point is that the body is all about knowledge and how to use it. The wisdom of the body is real but invis-ible. There is no reason to separate it from the wisdom of a philosopher, sage, or scientist. If intelligence is being used to solve a problem, that's a sign of consciousness. After all, these immune cells recognize strangers, act with a purpose, invent new defenses, read and receive messages, and interpret those messages accurately. Even in the brains of Alzheimer's patients, Rudy and his colleagues discovered that the pathological senile plaques are not just lethal chemical junk but are actually intended to protect the brain against infections from viruses, for example (see page 271 for more on this important finding). What other word makes sense except to say that all of this is conscious?

Medical research has performed a tremendous service by investigat-ing the body at a microscopic level, because in everyday life, what we mostly notice is the responses that occur on a macro level, like what you'd experience during a workout at the gym: sweating, heavier breath-ing, a faster heartbeat. Some adaptations are on the micro level, like the heightened process of bringing oxygen to your muscle cells and carrying waste products away, which also occurs in your workout. Medical science has devoted thousands of research hours to understand each adaptation in detail. But the whole-system approach is concerned with a much bigger mystery: How does the body *know* what to do?

Your body uses its intelligence on multiple fronts simultaneously, to keep itself balanced, strong, well defended, efficient, coordinated, and aware of everything happening in trillions of cells. Control by the host includes everything on this agenda. Moreover, each element is being handled in sync with every other element twenty-four hours a day. The rational mind isn't equipped to match the wisdom of the body. Consider what is needed, at a minimum, simply to get out of the body's way and stop undermining it.

In the Spirit of Cooperation
CHOICES THAT SUPPORT THE WISDOM OF THE BODY:

Reducing stress

Fighting against low-level chronic inflammation

Physical activity every day

Avoiding toxic air, food, and water

Eating a natural whole-foods diet

Getting good sleep every night

Being in a good mood

Taking time every day to be alone and quiet

Being centered in yourself, without undue distractions

Getting out of sympathetic overdrive, as covered in chapter 5

Approaching everyday challenges in a state of relaxed alertness

There are no surprises here, but we want to make two important points. First, the adaptive mechanisms in your body are preset to each of these actions by default. Your cooperation increases the bodymind's status on all fronts; your inaction decreases the bodymind's status on all fronts. A good night's sleep doesn't look like it has anything to do with immunity to catching a cold, the quickness of muscle response, the rhythm of eating and satiation, and not gaining weight. But on a holistic basis, getting a good night's sleep affects all of these things.

The second point, which follows from the first, is that you can't pick and choose to do one thing for a while and then move on to another. Your body operates on every front all the time. While you are focused on whether to buy organic spinach in the supermarket or finding time to make it to the gym, anything you leave out must still be handled at the level of your cells.

A natural response to all of this is "I can't do everything at once." Quite true, and that's been the big failing of holistic health—no one can actually encompass the whole bodymind. You do one thing and at the same time you leave something else out. Overcoming this obstacle is where the wisdom of mind must enter, not with laundry lists of good intentions but with a truly holistic way to increase the wisdom of the body.

Britt's Story: The Beginning of Wisdom

Britt is a beautiful Swedish woman with flowing blond hair; she looks much younger than her actual age of forty-eight. Until a few years ago, anyone looking at her life would conclude that Britt has been incredibly fortunate. Besides her physical attractiveness, she had a fulfilling family life, married to Poul, a successful private investor who had emigrated to America in his twenties (Poul had fallen in love with Britt while living in the United States and had divorced his first wife to be with her). Together they have three grown children, now away from home, who all worked hard to become educated and productive. Poul has been a devoted father, and their two children, now in late adolescence, have turned out to be happy and successful in school.

Then five years ago, without warning, when the whole family was gathered for Thanksgiving, Poul announced that he was moving out. "I don't love your mother anymore," he stated flatly. "It will be good for the whole family if she and I separate."

It wasn't just the shocking news that devastated Britt. "He made his announcement in front of the kids, not in private with me. And he acted so calm and certain."

Tears and wrangling followed. The children took sides, the two girls blaming Britt for not keeping their dad happy, the son wanting to protect her. But Poul was adamant. He had already rented an apartment nearby, and to his wife's utter amazement, he suggested that they all continue as before, being a family and doing everything together as before, only with him living elsewhere.

For the first month or two Britt went along. She was successful in her own right, working with a public relations firm. "I couldn't just drop everything and fall apart," she said. So Poul got his way. He moved out of the house but dropped in whenever he wanted to have dinner or to see the kids. When Britt demanded an explanation for his change of attitude, he revealed that he had begun to mistrust her. On a business trip several years before, she hadn't answered her hotel-room phone late one night. Poul was sure there were other men.

Despite her resolve not to fall apart, Britt began to feel increasingly anxious, and what triggered her anxiety the most was something quite basic, being alone. Finally, unable to sleep, feeling afraid at night, and not knowing where else to turn, she consulted a psychotherapist. He arranged for a prescription for tranquilizers while at the same time asking Britt if she had any idea why being alone made her so anxious. She shook her head and agreed to return for further sessions—taking a pill, she knew, wasn't the answer.

Over the course of the next few months, a pattern emerged. Britt had sacrificed herself for twenty years being the perfect wife, mother, and career woman. The burden of being a superwoman hadn't bothered her; in fact, she was proud of her success. But the therapist pointed out something that stunned her.

"You gave too much of yourself away," he said.

"What does that mean?" she asked.

"You gave yourself away by putting everyone else's needs ahead of yours," he replied.

Britt almost said, "That's what women do," but instead she reflected for a moment. "Everything I did created a loving family. At Christmas or on my birthday, everyone told me I was the center around which they revolved, the guiding star."

She started to cry, and there was no mystery to that. Having occupied her place at the center of her family life, Poul had undermined her security by telling her that he no longer loved her. It made her role seem hollow.

"You adapted to another person," the therapist said, "which happens in every marriage, and should happen. But it went one way. Your husband dictated how things ran. He held all the power. He made the critical decisions. Feeling totally in control, he could move out knowing that you would give in."

There was a lot more talk about what Britt had given away over the years, including self-esteem, dignity, and the right to make her own decisions. Her story could be about a woman getting her life back on track after a devastating breakup—which was a major part of it—but one day she asked her therapist a critical question: How do you get back the self you gave away?

The therapist was surprised. "Are you seriously interested in that?" he asked. In his counseling of couples going through a divorce, the focus was usually on getting back at each other, overcoming feelings of betrayal and bitterness, and emotional recovery. These things take years, and not everyone makes it through the recovery process in good shape emotionally.

"You said I gave myself away," Britt insisted. "I want back what's mine."

Britt wanted to become more whole, to reclaim an inner life that wasn't dependent on giving up her power, self-esteem, and freedom to have her own opinions and beliefs. She was looking for healing at the level of the self. But which self? There are several possible versions of the self that you can identify with, and how your life turns out depends on which one you reinforce. "I" is more elusive than most people realize. Let's look at four possible choices.

The outward self: This is the social persona, which you identify with if your focus is on socially approved things like money, career, the right neighborhood, an impressive house, and so forth. "I" is attached to labels that relate to those things, so that "WASP surgeon with a Park Avenue practice, a socialite wife, and a major portfolio" defines a very different self than "Latina working-class single mother living on food stamps."

The private self: This is who you are behind closed doors. The private self identifies with feelings and relationships. The values that matter most include a happy marriage, satisfying sex life, children to love and be proud of, and so forth. On the downside are the private trials and miseries that come into every life. "I" is attached to the hopes and fears of everyday existence, which for some people means an existence of insecurity, anxiety, depression, and dashed hopes that seems inescapable.

The unconscious self: This is the self we do not know in waking life. It is governed by instincts and drives that most of us don't want to bring to light. At its most menacing, the unconscious self has been called "the shadow," where the worst human traits of anger, violence, envy, revenge, and deep-seated existential fear reside. In relation to the shadow self, one might attempt to keep the dark side of the unconscious self-hidden or try to convert it to the light. Artists, musicians, and poets do the latter. They approach the unconscious self not as a fearful domain but as a source of creativity waiting to be born.

The higher self: This is the self that aspires to rise above everyday conflicts and confusion. Experience tells us that the other versions of the self—the outward, private, and unconscious self—are constantly in conflict. This is why civilization is so discontented, to use Freud's term. Eruptions from the unconscious bring war, crime, and violence. Private misery overshadows public success. The arts point to immense possibilities for creativity, but too few people are able to take advantage of them. In the world's wisdom traditions, the struggle between so many conflicts can't be won at the level of struggle. "I" must surrender every claim of the ego, whether public or private, to seek a higher state of consciousness.

Britt wasn't unusual in finding herself mired in struggle. A crisis like the one her husband threw her into creates heightened confusion and struggle, but everyday life masks the same conflicted situation. Nor is Britt unusual in the choices she made. In exchange for an outward self that looked perfect to the outside world, she gave up her power over the private, unconscious, and higher self. What's unusual is how quickly she came to this realization after her husband moved out.

Her therapist was very encouraging. "Whatever you gave away, you can take back," he said. "It's a return journey to pick up the pieces you dropped along the way."

The world's wisdom traditions, which we define simply as the traditions of higher consciousness, agree. Wisdom begins by recognizing that the bodymind isn't just about cells, tissues, and organs, nor is it just about thoughts, feelings, and sensations. Instead, the bodymind is about unity between body, mind, and spirit. If you fall in love, there is a corresponding biology of love created in your body. Likewise there's a biology of anxiety, a biology of depression, a biology of happiness. The whole-system approach rests upon this fact, but it's hard to grasp the whole truth: there is a shifting biology that meets the needs of every moment. Your cells know what to do in any situation, which counts as the most astonishing level of intelligence in nature.

The "I" you identify with is like a magnifying glass gathering the sun's rays to a point. Your "I" interprets every experience and makes it personal. "I" is a bundle of hopes, fears, wishes, and dreams. "I" harbors memories no one else has, and in the compartments of memory are stored habits, beliefs, old traumas, and past conditioning. This multiplicity is bewildering, which is why the teaching of "Know thyself" is actually the point of being alive—until you know where "I" came from, you cannot discover who you really are.

Britt took the "journey of return" seriously. Despite all the advantages of her external life, she couldn't do the most basic thing, which is to be alone. Without a life of keeping busy and caring for everyone else, "I" was a terrifying prospect to her. This implies a great deal of healing in the unconscious, where the demons lurk but also the wounded child. Britt's journey over the next five years went through roughly the following stages:

Overcoming her anxiety. Britt relied on tranquilizers at first but weaned herself off through therapy and, more important, by taking up meditation and yoga.

Learning self-reliance. Britt told Poul he couldn't come around anymore pretending that family life was normal. (It quickly came out that he had had a girlfriend on the side.) She moved toward divorce on her own terms and at her own speed, taking fully two years before she felt ready to stand on her own.

Getting back into relationship. Britt began to date, a strange experience for a woman who hadn't been on a date in twenty-five years. She discovered that she wanted to be happy again, and at that stage she took up dancing, a love of hers from her teens, and steadily developed friends outside the couples she and Poul had known together.

Finding a spiritual path. Britt took her meditation practice more and more seriously, looking far beyond issues of stress, relaxation, and health. She had deeply absorbed the lessons of what happens when you give yourself away—you become unconscious. Beneath her anxiousness about being alone lay a kind of numbness. Her active, busy, successful life had soaked up all the energy. Deep down nothing was moving. The inner woman was stuck and had been that way for years.

All of us resemble Britt, not in the particulars of her story but in the return journey we must make if we want to heal. "I" must reawaken to the possibility of a vibrant existence once again, where the light of awareness is the true healer. When you lead a conscious life, the following experiences are real possibilities, and they can unfold at any time:

The Riches of a Conscious Life
ALL THE POSSIBILITIES YOU COULD HAVE TODAY:

You give someone a helping hand.

You notice something beautiful.

You do or say something kind.

You offer yourself in service to someone in need.

You smile with appreciation.

You forgive a slight.

You make another person laugh.

You have a fresh new idea.

You find the solution to a problem.

You feel a close bond with another person.

You meditate.

You take time out to be alone, valuing your private time.

You lift someone else's spirits.

You are playful and take time to play.

You walk outside in nature and feel refreshed.

You engage in physical activity that's invigorating.

You respect someone else's boundaries without being asked.

You feel light and buoyant.

You feel spiritually uplifted.

There is a moment of pure joy.

You cherish another person.

We don't have to explain why these experiences are desirable; the fact that each one of them creates a moment of happiness is obvious. The real question is how to create them. Each version of the self has a different point of view, each its own goals.

The outward self doesn't look inward, because its goal is to achieve happiness with outward success and the accumulation of money, possessions, status, and so forth. The private self does go inward, feeling the ups and downs of emotions. It wants to achieve happiness by having more pleasure than pain. Perfect happiness would be a state of constant pleasantness. We all know this is unrealistic and will never be attained. Yet most people expend a great deal of time and energy doing what they can to have more positivity in their lives than negativity, however you want to define those terms.

The private self can experience some of the riches of a conscious life, since our efforts to be loving or kind, for example, are often rooted in our emotional life. It feels good to be kind and loving; therefore, most of us enjoy the experience. But there are limits. The private self is selfish and insecure. Given a choice between its own happiness and someone else's, it chooses the former. If a loved one withdraws their love, as Poul withdrew his from Britt, the private self experiences a sense of loss and pain. The prospects for a pleasant existence fly out the window, at least for a time.

The unconscious self is a mysterious part of the psyche, a hidden region most people fear. Who knows what its purpose is, or what it needs to be happy? The greatest conflict in modern psychology was over this very issue. Sigmund Freud came to believe that the unconscious was the domain of the *Id*, a primal force that is untamed. Id isn't held back by guilt or shame; social rules don't touch it. A two-year-old throwing a tantrum in a grocery store might be a good example of pure Id breaking out. The child feels no restraint in displaying his anger and doesn't care who he hurts or embarrasses. The tantrum, like the Id in general, isn't immoral or even selfish. Id is simply ungoverned, and we often fear its lawlessness, just as Freud traced every dark force—hatred, aggression, sexual appetite, the lure of death and violence—back to the unconscious.

But his most famous disciple, the Swiss psychologist Carl Jung, strongly disagreed and eventually split off from Freud entirely. Their disagreements are complicated, but one of the essential issues was Jung's insistence that the unconscious didn't simply hold dark forces. It held dozens of patterns or templates of behavior that Jung labeled archetypes. The human race shared these patterns in its *collective unconscious*. As proof, Jung pointed to the way every society had heroes, myths, gods, journeys into the light, questing, fixed models of masculinity and femininity, and much more. He conceded that the unconscious could erupt in war and violence, but for him this was the expression of one archetype (such as Mars, the Roman god of war). Yet in the scheme of archetypes, there is also Venus, the Roman goddess of love.

Jung worked closely with Freud from 1907 to 1913, but over that time their relationship became more and more tumultuous. After his split with Freud, Jung began work on what many consider his masterpiece, *The Red Book*, or *Liber Novus* (the "New Book"). The book was largely based on the vivid and often disturbing nightly dreams Jung had while he was serving as an officer in the Swiss army. Many believe that Jung was practicing lucid dreaming, or waking up in dreams, which he would journal with beautiful calligraphic narratives and detailed artwork of impressive quality. He believed that his dreams were a window into the background activity of his unconscious, which he chronicled for sixteen years.

Before he died in 1961, Jung told an interviewer:

[W]hen I pursued the inner images [was] the most important time of my life. Everything else is to be derived from this. My entire life consisted in elaborating what had burst forth from the unconscious and flooded me like an enigmatic stream and threatened to break me. That was the stuff and material for more than only one life. Everything later was merely the outer classification, scientific elaboration, and the integration into life. But the numinous beginning, which contained everything, was then.

The Red Book in its handwritten leather-bound form dates from 1915 to 1930, but it was only published in 2000. (Free facsimiles, including Jung's elaborate illustrations, can be found online in PDF format.) Various commentators, like its translator, Sonu Shamdasani, look upon the book as Jung's tortuous quest to salvage his soul through an inner dialogue with his *unconscious*, what he called the "spirit of the depths." It's sometimes contended that *The Red Book* was the result of a psychotic break after Jung's split with Freud. Sympathetic supporters hold that Jung boldly faced his own psychosis head-on, confronting what he found deep inside his psyche via his dreams, and emerging stronger and more whole for doing so.

This essential conflict between Freud's Id and Jung's archetypes had a powerful influence over the field of psychology for many decades, and even today hasn't been settled. Perhaps it never will be, yet for each person in everyday life, there are experiences of desire, appetite, anger, and the temptation of violence that cause disruptions no one wants to experience. The outward self gains a lot of its power by pushing down the unconscious self. Every time you go to work, you put on your outward self, and so does everyone around you. The unacceptable alternatives, like sexual harassment and open hostility, are kept in check as much as possible. The upshot of this rather lengthy description is that the unconscious self isn't opened up in normal life. If, as Jung says, we could find something beautiful and gratifying by exploring our unconscious, few dare to open that door.

What are we left with, then? Only the higher self has access to the rich experiences we call the conscious life. Its goal is to live in the light of awareness, which isn't the same as being pleasant all the time. Awareness is unfiltered and free. Its openness to every experience represents a leap of faith. But those who have taken the leap, including the world's sages, saints, and spiritual teachers in every culture, declare that the higher self is real—in fact, it is the only real self. The other versions of the self are unreliable. They make false promises, suffer from insecurity, fear loss of control, give shelter to hidden demons, and ultimately cannot reach any permanent state of happiness.

Britt discovered this by going through a personal crisis. She is one among countless people who have chosen to walk a different path, to discover for themselves if the higher self can be found. She is on a healing journey because that is where she found herself as the crisis unfolded. But it doesn't take a crisis to initiate this journey. On any given day, the one thing we cannot live without—a self—is shifting and unreliable. We may not be aware of it, but we are constantly changing our loyalties. The outward self claims us at work or enjoying ourselves at a party or buying a new house. The private self claims us in matters of the heart, in moments of depression and anxiety, and in our family life. The unconscious self does whatever it wants to, and hard as we try to keep it at bay, everyone knows the experience of sexual appetite, raging fury, and nightmares— perhaps nightmares are our purest encounters with the dark side of the unconscious.

The instability and unpredictability of the self, the "I" we take for granted, poses the final challenge to healing. The thing that looks so simple—getting out of the way so that the bodymind can heal itself— turns out to be very difficult. The wisdom of the body is incredible, but we undermine it through the stress and unpredictability of everyday existence. Instead of having a healthy relationship to the self, we constantly question who we are. We plunge into situations we can't handle and relationships filled with hidden conflict. Our attempts at self-control are temporary and only partially effective. If we succeed in being perfectly controlled, the cost is paid in the negative emotions we push down out of sight.

In short, the situation is messy. To be a wise healer, you have to solve the problems created by "I" and its many problems. But how can "I" be part of the solution at the same time that it is the source of so much harm? Asking the self to heal itself is rather like asking a surgeon to apply a scalpel to remove his own appendix. Needless to say, most people never resolve this paradox—they live year after year with an "I" that gets along as best it can. Experiences come and go. Good things happen one day, bad things the next. In the end, people arrive at a state of health and well-being that has little rhyme or reason. They're stuck with what they've got.

In the next chapter we'll see if this haphazard outcome can be changed. There has to be a better way and, in fact, there is.

10

The End of Suffering

*If the healing self could bring suffering to an end, it would seem mi-*raculous. Every life brings some pain, and the mental side of pain, which is suffering, comes along with that. No one escapes the inner drama of the psyche, no matter how happy their life is on the surface. (In the last chapter we talked about how Jung openly embraced his own inner drama.) We've based our whole-system approach on leading a healing lifestyle, which means a lifestyle in which consciousness comes first. Essentially, what you aren't aware of, you cannot heal. Juries often give massive awards for pain and suffering, but the two aren't the same thing. You can have acute physical pain and adapt to it psychologically, which reduces the quotient of suffering much more than for someone who cannot adapt.

Simply anticipating bad events in the future creates a level of stress in the bodymind that can create physical pain (e.g., a big meeting with a boss about your performance could lead to chest pain, headaches, lower-back pain, upset digestion). When these symptoms appear, some people will also feel mental suffering, such as fear, anxiety, and depression. But others won't. In other words, suffering is more personal and elusive than physical pain, which everyone notices. It would be very unhelpful if knowing that you are suffering harmed you in comparison with someone else who goes into denial. Unfortunately, this is a commonly held belief.

"What you don't know can't hurt you" causes its own hurt over the long run. Fear is often driven by subconscious memories of pain and suffering in the past, which can create even more suffering in the future.

Suffering isn't a topic that's easy for most people to talk about, but there are good data on happiness around the world, which would correlate to levels of suffering. The Gallup organization, best known for political polling, also gathers data worldwide on how happy people are. This is done one of two ways, either by asking responders to self-rate how happy they are or by posing a simple question: On the day before the interview, did the person laugh and smile a lot? Gallup's highest happiness ranking is "thriving," and in the United States, by their current estimates, only 51 percent of people say that they are thriving, placing us 14th among the 142 countries Gallup measures. Only 4 percent of Americans respond that they are suffering, while 45 percent report that they are struggling. (By contrast, in India, which ranks 127th in happiness, only 8 percent are thriving and 28 percent are suffering, leaving a large majority to struggle.)

It may be, as some experts believe, that people overrate their happiness if you consider the hidden or unspoken causes for unhappiness. Approximately one in five Americans will suffer a bout of severe depression in their lives. Domestic abuse is notoriously underreported and belongs among the factors that are not addressed in mainstream medicine. Even in the world's two happiest countries, Denmark and Norway, where 68 percent of people report that they are thriving, 30 percent are struggling. By implication, there are millions of people in the United States who urgently need to get out of suffering, whether that means ending an abusive relationship or leaving a job that causes one to get sick.

At the end of the last chapter we introduced a new possibility, the higher self, which expands how powerful consciousness can be. *Higher* brings in a spiritual connotation, which we need to explain before going any further. The separation between body and mind is artificial, and medical science strongly supports merging them into the bodymind. That much seems settled. "Higher" consciousness would appear to cross the line into the domain of God, spirit, and the soul, which medical science has nothing to do with. Every hospital has a chaplain, yet he doesn't stand in the operating room beside the surgeons.

If the healing self is going to bring an end to suffering, another boundary has to crumble, because meditation research, now a completely accepted field, employs a spiritual practice. It might seem peculiar that popular television personalities like Dr. Oz and Dr. Phil would be joined by Dr. Buddha, but that's entirely plausible. The Buddha offered a path to end suffering based on consciousness, not an appeal to God, spirit, or the soul. Meditation is consciousness-based medicine. Whatever happens when you are meditating (or praying, doing yoga, being mindful, etc.) registers in cellular activity, first in the brain, then farther down the line in the rest of the body.

This leads to a simple but powerful conclusion: the end of suffering is a consciousness solution to a consciousness problem. No one suffers because they are in pain. Suffering is an interpretation based on all the things we've been discussing: beliefs, habits, old conditioning, and the struggle between acting mindfully or mindlessly. If you change your interpretation, your degree of suffering will change. The higher self represents a major change at the level of "Who am I?" When you identify with the higher self, you discover the path out of suffering, because you discover inside yourself that the following things are true:

There's a level of awareness that experiences no suffering. Painful experiences register but do not linger as suffering.

Physical pain exists as a sensation, but this is a signal for healing, not a curse.

The source of suffering is the same as the source of healing: your state of awareness. We aren't denying the benefits of pain research and the need to relieve physical pain. The first question the doctor asks the patient is "Where does it hurt?" The aim on both sides is to become pain-free. Our goal in this chapter is to become free of suffering, which can only occur at the level of consciousness. (Dr. Buddha would say the same.)

The Paradox of Pain

Few physicians specialize in pain per se. To the typical MD, pain is something to get rid of, not to understand. But if you try to understand the mechanics of physical pain, the picture isn't simple. Sometimes pain is like a rock in your shoe, which you immediately take out to relieve your discomfort, or a toothache that sends you rushing to the dentist. But at other times physical pain isn't immediate or easily fixable. In fact, a pain that signals serious long-term damage somewhere in your body is often the last symptom to appear. Many of the most widespread disorders that medicine still wants to cure, such as heart disease and cancer, can go for years without sending the signal of pain, by which time the possibility of preventing them may be long past.

Consider aging, which typically brings its share of aches and pains. They don't have to convert into a state of suffering, however, and when they do, the linchpin is a person's beliefs. In a society where billions of dollars are spent on pain relievers while suffering is a difficult topic that most people refuse to confront, beliefs exert hidden power. A typical chain of reasoning might be the following:

Pain creates suffering.

The more intense the pain, the greater the suffering.

As we age, we can expect to get increased physical aches and pains.

Therefore, aging brings increased suffering.

These are beliefs that have a shaky relationship to reality, but if you hold them strongly enough, the bodymind converts them into *your* reality. To begin with, people need to revise their belief that pain equates with suffering. Pain by itself is often something we can work through and ignore. The "no pain, no gain" ethos in sports is a prime example. Marathon runners volunteer to undergo considerable pain to achieve the bigger goal of victory. The desire to win can become so all-important that a severe, life-threatening condition is tolerated, such as repeated brain trauma in boxing, football, and rugby, all the way down to junior leagues, where a child's future health may be at stake.

In society's panic to make pain the enemy, listening to your pain and discomfort, which is the body's primary reason for sending pain signals, comes second or not at all. This indicates that our priorities are confused. A life without the body's pain signals turns out to be far from fortunate. There is a genetic condition that deprives certain people of any sensation of pain, and those who suffer from it find everyday life quite threatening.

One patient born with this genetic condition, known as CIP (congenital insensitivity to pain), is Jason Breck, as we'll call him. The condition is exceedingly rare, with only about twenty documented cases in the medical literature. Breck's parents discovered his condition when as a toddler he bit off half his tongue. Interviewed as an adult, Breck recalls, "One incident I remember is breaking my foot on my birthday. It was swollen and bruised, so I took some duct tape and taped my foot, put on a boot, and just continued with my day." For some time there was doubt that such a condition existed, but now it's known that CIP results from a mutation in a single gene (SCN9A), and even more startling, that one molecule is responsible for controlling pain. The mechanism is linked to the fact that SCN9A sits on the neurons that cause the sensation of pain.

Since there is also generally an insensitivity to temperature, people with CIP are surrounded by dangers the rest of us don't experience. "You have to be hypervigilant in your day-to-day tasks," Breck says, "to avoid causing serious injury to yourself." Without the signal of pain, other strategies must be developed to tell you that you are injured. The sensation of touch usually isn't damaged, so sensing pressure or a sudden knock serves as a cue. But this also poses a danger. As a young child Breck liked to bang his head against the wall to feel the vibration, which he enjoyed. Children with CIP therefore need to wear helmets to avoid this dangerous behavior. (In Breck's case, he is able to sense temperature but has no sense of smell, another threat since, for example, he wouldn't smell smoke from a house fire.)

It took the inheritance of the SCN9A mutation from both his father and mother to create Breck's condition, so the genetic probabilities are extremely low. The genetic clue also could turn into a powerful pain-killer. If the signals from normal SCN9A could be temporarily blocked after surgery or severe injury, for example, the pain relief would be total and, at best case, without any side effects. Even more starkly, a sizable percentage of terminal patients who want to have assisted suicide are in unbearable chronic pain that the most powerful narcotics can't relieve. For them a genetic treatment may be their only hope.

If you look at the wider picture, however, what this example is about is the paradox of pain. Because it evolved as a sensation to serve and preserve us, but can harm us at the same time, pain is one of the most elusive things to handle in our lives. There's no getting around the fact that pain alone doesn't cause suffering. Not just your beliefs matter here. In a telling 2013 study, Antoine Lutz and his colleagues wanted to test whether being open to the experience of pain—in other words, mindful—works better than the usual tactic of avoiding pain and feeling anxious before the pain occurs.

As the researchers note, very little is known about how mindfulness affects the brain activity associated with pain. As subjects they took a group of "expert meditators" who had been doing their practice for more than 10,000 hours and used fMRI brain scans to look at brain activity associated with anticipating pain, experiencing pain, and getting used to pain. When subjected to a painful stimulus, the expert meditators felt the same intensity of pain as novice meditators, but they reported that it was less unpleasant—that is, they suffered less. In terms of what was happening in the expert meditators' brains, the researchers commented that "this difference was associated with enhanced activity in the dorsal anterior insula (aI), and the anterior mid-cingulate cortex (aMCC), the so-called salience network. In neuroscience, *saliency* refers to how much a thing stands out compared with its neighbors.

But why would lifelong meditators notice pain more quickly and yet suffer from it less? The key is that their baseline for pain was lower than the control group; they didn't have as much anticipation of being hurt or the anxiety that comes with it. Then when pain was felt, they registered it quickly, and they got used to it more quickly. This is the rather technical story told by brain scans. But it accords with subjective reports from meditators that they feel calm, centered, and at peace.

Our view of these findings is simple: consciousness can intervene to reduce suffering, even when the level of physical pain doesn't change. What do we learn from this? To be healed is to be free from suffering, and if this ideal cannot be readily attained, each of us should try to get as close as possible. Let's look at someone who set out to achieve the goal.

Darren's Story: Change and Renewal

Darren, who is forty-five, is married, and lives in Colorado, didn't set out to revamp his identity. Yet somehow it happened, and the results have been dramatic—classmates who knew him in college are astonished at the changes.

"I didn't come from a rough background or a difficult family," Darren says. "I felt completely normal—a brash, competitive kid heading for a career as a lawyer or doctor. Something good that paid really well."

With this vague goal in mind, Darren felt well equipped to succeed, although behind his back others found him too aggressive and even arrogant. Classmates let him have his way, not because they necessarily liked him, but because he dug in his heels or retaliated if anyone opposed him.

He smiles with regret. "I know I was a jerk, and as far as anyone could predict, that wasn't going to change."

But then a family tragedy occurred. His younger brother enlisted in the army, went to fight overseas, and didn't return.

"I rushed home to be with my parents," Darren says, "and as devastated as they were, I felt numb. I couldn't even cry. One day, two soldiers arrived at the door to deliver my brother's posthumous medals for bravery. My father barely said a word to them, but when they left he opened the box and said, 'Take a look at what your kid brother died for.'"

Perhaps it was significant that this major disruption in Darren's life happened when he was twenty and he was still malleable. At a time when identity enters a crisis for most young people, he was more than badly shaken.

"I started to hate myself—that's not too strong a word. I began drinking too much and playing video games until three in the morning, but nothing really cut through the guilt for more than a few hours. I was supposed to protect my little brother, yet I had barely paid attention to him. I lay awake every night wondering if I could have stopped him from enlisting, until I realized that I didn't even know his motivations. Was he out of options? Did he have a deep streak of patriotism?"

A period of self-examination began to unfold. Instead of going directly to law school or medical school, Darren took time off, supporting himself with casual work like house painting. He didn't find a solid, stable relationship, and after a year or two, he stopped dating.

"I had figured out something," he says. "If I didn't work on myself, only two options lay ahead. Either I would be stuck with some unbearable emotional baggage, or I could just pretend to myself that I was okay—then what?"

Over the next five years Darren took the exceptional step of going inward and staying there, investigating who he was. "I wasn't qualified to psychoanalyze myself, but that wasn't it, really. I just wanted to be able to feel right with myself again, and to get there, I had to deal with the fact that I had turned into the kind of person I never wanted to be—not just a jerk but a man with no inner life."

Darren's decision isn't unique—countless people have decided, for hundreds of reasons, to step away from society and walk the inner path. Whether they consider this path spiritual or healing, walking the inner path requires a new kind of awareness, one that few people are prepared for. How do you rearrange your inner life, with its clutter of old memories, habits, wounds, and conditioning? Everything "in here" is invisible. Unwanted emotions like fear and depression wander at will in times of crisis.

Despite these difficulties, Darren kept being motivated by one thing: self-renewal: "I refused to buy into the idea that I was a finished product, that I'd show up at a class reunion twenty years from now and everyone would say, 'You haven't changed a bit.' To me, that would be a terrible way to turn out."

Seeking self-renewal is a conscious decision, and it's never once and for all. Renewal at the level of the cell is a constant and virtually automatic process, and so is self-renewal. The noted spiritual teacher Jiddu Krishnamurti once made a provocative remark about meditation. People set aside a certain period of the day to meditate, he said, but they don't realize that real meditation is twenty-four hours a day. The same is true for healing. Cells don't see this round-the-clock duty as an obstacle.

On the personal level, however, round-the-clock healing seems impossible. But if you look more closely, to heal twenty-four hours a day isn't like choosing to watch TV or bounce a basketball twenty-four hours a day. Healing is more like breathing, a life-sustaining process that works automatically yet can also be augmented (by the breathing exercises in yoga, for example). Because healing is an automatic process, you are totally immersed in it already. So what was Darren really choosing to do? How could he know whether his project of changing himself would work in the first place?

He began with a belief that everyone needs to adopt: There is no fixed "I" or self. From this moment onward, you will never be the same person again. Therefore, it is futile to hold on to "I" like a raft in a storm at sea. The "I" self *is* the storm. Everyone is tossed about by every kind of external and internal force, and as all of this turmoil crashes around, the bodymind bends with the currents of change. Your conscious mind cannot possibly keep track of this turmoil. We're incredibly fortunate that evolution developed the healing response to such perfection that living on automatic pilot saves us from being damaged by the changes that assault us.

What Darren and millions of others have discovered is that evolution can take a new direction—it can become conscious. This puts healing in a different light. Instead of making positive lifestyle choices your chief focus (beneficial as those choices are), you immerse yourself in the healing process, becoming the healing process. It becomes your goal to evolve into a "higher healer," if we can put it that way. Here is what this involves.

How Higher Healing Works

You place a high value on happiness.

You live from a stable center core.

You stop struggling and resisting.

You seek harmony by serving as an example to others, not trying to control them.

You choose resonance with others over dissonance.

You remain open to what's happening here and now.

You have a vision of the best life you can lead, based on values larger than yourself.

You pay attention to subtle signs of distress and discomfort.

You undo the damage of the past.

You look forward to the future with optimism.

You enjoy being constantly in process.

These are the traits of conscious evolution. By setting as your goal that you want to grow and evolve every day, you are entering into an enlightened partnership with everything happening to you, holding no judgment against yourself or what your life is bringing to you. Since every process in the bodymind is self-organizing and self-renewing, the most evolved way to live is to let yourself unfold naturally. Words like *flow* and *surrender* come to mind, yet they barely touch upon the reality of dedicating yourself to constant renewal. Lao-Tzu, the father of Taoism, teaches that we must endure what life throws at us like reeds in the wind. If we bend and yield to the pressures of life, allowing the natural course of events to sculpt us, we endure. If we insist on remaining firm and upright, we snap.

"Looking back, I see my path as one thing," Darren says. "Can life be trusted? Can it take care of itself? I put it that way because my brother's death brought up a deep streak of distrust. Life had kicked me in the teeth—now what? Most people take the blow, try to survive it, and then piece together their version of a normal life. But they never solve the underlying question. Can you really trust what life is going to bring you? If not, you better build a wall around yourself and hunker down for the worst." You can call this one man's philosophy of life, but the issue runs much deeper.

The Mysterious "I"

As we've noted, none of us has a fixed self. We are constantly changing. We often observe this when we try to soothe an infant who is teething or need to buy new outfits for a teenage son or daughter because they've outgrown their old clothes that fit just a week ago. In reality, nothing about you is exactly the same as it was yesterday. Who are you, then? "I'm a process" sounds strange to most people, but let's examine the science that supports this answer.

In a 2016 TED talk, Moshe Szyf, a prominent geneticist at McGill University, outlined a fascinating study that involves rats and how they are mothered. A "good mother" rat shows her goodness by licking her newborn offspring much more than a "bad mother" rat, who neglects this task or does it half-heartedly. When they grow up, the pups lead very different lives—the ones who had good mothering are more relaxed, less stressed, and exhibit different sexual behavior than the rats from bad mothers. Normally, a geneticist would say that there is a specific gene being passed on that determines what kind of mother a rat will turn out to be.

But Szyf is a specialist in epigenetics, which studies how the genes we're born with are affected by life experiences—the epigenome is all the factors that control the activity (expression) of our genes. It includes chemical modifications of our DNA and the proteins, known as *histones*, that sheathe and surround it. DNA and its protein sheath get chemically imprinted by experience and play a major role in turning genes off and on, up and down. (This process was a major focus of our previous book, *Super Genes*.) Over ten years, Szyf and his colleagues explored what might happen if a baby rat from a bad mother was given to a good mother for its care, and the same in reverse for baby rats from good mothers. What the researchers discovered is that a significant number of chemical pathways were changed, and this supported what they could see—a good mother rat could turn her adopted baby into a relaxed, unstressed adult—in other words, the experience of being raised well overcame the inheritance from a bad mother. The reverse was also true. Even though a baby rat was descended from a line of good mothers, this good inheritance could be reversed by having the rat be adopted by a bad mother.

Szyf goes on to speculate about more than nature versus nurture in rats. What is it about the way a human baby is raised that fixes its ideas about how life works? Szyf points to a baby raised in Stockholm, where winter days are cold, bright, and very short, compared with a baby raised in a tribe in Brazil, where days are of equal length and hot. A baby whose system received these different inputs would go on, he speculates, to expect life to be different based on infant experience. There would be expectations about other important things, such as the abundance or scarcity of food, a sense of danger or safety, or an easy or hard survival in general. Thus, Szyf declares, evolution has taught our old fixed DNA to adapt dynamically to all kinds of environments—a major clue to something we have been saying, that you belong to a species that is the most adaptive on Earth.

Now we reach a crossroads. Is this early set of imprints the key to health and disease? The sword cuts both ways. The same imprint could help or hurt you in later life, and there's no predicting which will occur. Let's say child A is raised to feel safe and protected, and he grows up holding this belief inside, while child B is imprinted with the idea that life is unsafe and unpredictable. We might say that child A is going to get along more happily than child B. But what if danger looms on the horizon, like the first AIDS case or the emergence of a Hitler or a Stalin? The child who assumes by default that everything will work out well in a benign, safe world could be tragically unprepared as an adult facing a looming threat, while the child reared to assume by default that worst-case scenarios must be confronted could turn into the only survivor of the two.

Szyf arrives at a groundbreaking conclusion. Thanks to genetic advances, one can now see where exactly the entire genome has been marked by good or bad mothering. He points to a study of monkeys and mothering where one baby had a real mother while another had a surrogate doll. Many genes looked different between the two groups, beginning as early as fourteen days after the infant's birth. "This signals the whole way life is going to look like when you become an adult," Szyf declares. "Stress rearranges the entire genome." How early do all of these differences show up? The question has a bearing on experiences in early infancy. For example, when caring for an infant, parents can choose to let the baby cry in attempts to train it to sleep through the night, the so-called Ferber method. Or like Rudy and his wife, Dora, with their baby daughter, Lyla, they can attend to the crying every time it happens.

In the latter case, it is tougher on the parent, but the earliest neural networks and genetic imprinting in the infant will program a message that can last for life, that the world is a good and secure place. Of course, the world will turn out to be full of challenges and letdowns, but this positive imprinting very early on goes a long way to promoting healing over suffering all through life.

We may already know—and be programmed to accept—our place in the world from the moment of birth. Animals obey such programming instinctively. For example, monkeys always arrange themselves in a social hierarchy, with a dominant monkey at the top and the lowest monkey at the bottom. The differences in their genomes are already present when they come out of the womb, which in human terms may mean that belonging to a disadvantaged background would imprint a baby from the first day. What makes this possibility more distressing came out in a study that went back to an ice storm in 1998 that wiped out power in the dead of winter for all of Quebec. This event was more stressful for some people than others, and among the population were pregnant mothers.

Following their infants for fifteen years, the developmental psychologist Suzanne King found that children of mothers who went through highly stressful experiences during and after the ice storm suffered higher incidences of autism, metabolic disorders, and autoimmune diseases. Of course, one cannot assume a cause-and-effect relationship here. Yet there are many roads leading away from these and similar studies, such as the discovery that single events occurring at a specific time during pregnancy can affect the development of the fetus. But the bigger picture is about how unstable the self really is, even though we think that "I" remains the same from year to year.

The longest study pertaining to the instability of self comes from Scotland, where in 1947 schoolteachers were asked to rate their fourteen-year-old students on six personality characteristics: self-confidence, perseverance, stability of moods, conscientiousness, originality, and a desire to learn. A total of 1,208 students was involved, and in 2012 a follow-up study tracked down survivors who would agree to rate themselves on the same characteristics, a total of 174. To gain perspective, the survivors were also asked to find someone who knew them well and who would give them a separate rating. It's long been held in psychology that personality is stable, and there's a common bias that "people never change." But the Scottish study came to the opposite conclusion. Although there were similarities between the younger and older age groups, the "correlations suggested no significant stability of any of the 6 characteristics."

No one quite knows why previous studies indicated that personality remains stable over time. Mothers often say that they saw in their babies the sort of personality the infant would grow up to have. "You were a quiet baby, and you're quiet now" is a typical remark, or "You always wanted to get your own way, even when you were two." But it appears that time explains a good deal. The Scottish study is the longest on record, and by their seventies these people had "hardly any relationship at all" to their adolescent selves.

The opportunity to transform your sense of self has always been there. Just as important, a lifetime of experience is going to transform you anyway, and the more time you wait, the more you will be transformed without your knowledge and consent. We'd like to point out some basic conclusions:

Early experiences imprint a child much more than anyone thought in terms of genes, biology, and behavior.

In the mix of all these influences, each of us carries around a map of life that isn't our choice but an imprint.

We can change this imprint by choosing the beliefs, behaviors, and interpretations we actually want. In other words, unconscious imprints can be undone once you set your mind to it—no other living creature is blessed with this possibility as far as we know.

There have been medical anecdotes about total changes of identity. In the 1960s, a brilliant but eccentric Scottish psychiatrist named R. D. Laing was in vogue. Laing reported the case of a young woman who had lapsed into a coma and who suddenly woke up. She knew herself by her own name, but a strange process of transformation had occurred. The girl had been shy and introverted before. Now the nurses treated her like a celebrity, the life of the party. They complimented her for how witty and charming she was. In short order, believing what she was told, the patient underwent a personality change, becoming the person that was being reflected back at her.

If we know that "I" can be dismantled piece by piece, either through brain injury or for psychological reasons, then the self is much less stable and trustworthy than anyone ever thought. Which leads us back to Darren. If it's true that a fourteen-year-old wouldn't recognize himself at sixty or seventy, Darren didn't want to. He drastically changed his former self because he couldn't live with the old "I" any longer. Now that he's in his forties, where did life take him after he turned his back on an unacceptable "I"?

"I do occasionally meet old friends from high school and college who say, 'You haven't changed a bit,'" he admits. "But I can laugh it off. I know they're just trying to be complimentary. If they really knew me, they'd be shocked, because the way I feel about myself isn't remotely like what I felt back then. I used to run away from myself. I had a voice in my head reminding me every day that I wasn't good enough. All of that's gone now.

"The voice in my head that was constantly judging me took a long time to go away. A thousand times, or ten thousand times, I would tell it, 'I don't need you anymore.' I used to be proud of how tough and hard I was, and that took a long time to turn around, too. But you can't be alive without feelings, and you can't feel unless you expose your vulnerability. I doubt one guy in a hundred faces that truth. I had to, because the thing that got me on the path, my horrible guilt over my brother's death, was too real to deny.

"On that front, my biggest lesson was that emotions can be a positive thing in your life. And a lot of other stuff began to shift. The whole big tangle about whether I was lovable and could love someone else—that could fill a book by itself. But if you lay every issue out in advance, you'd be paralyzed. I'm a believer in letting things unfold the way they want to. I don't fight or resist anything. When you aren't scared of yourself anymore, you aren't scared of your feelings, either, or of what other people say. You don't worry about the future or relive the past.

"Somewhere along the line, I wasn't escaping my pain. I shifted into another gear and became interested in what was going on with me. It was a project, almost like observing another person under the microscope. When the fear and judgment are gone, you begin to enjoy the project."

What exactly was this project?

"Self-discovery. No single term is good enough, but that will do," he says. "When you ask 'Who am I?' the answer goes through stages."

What stage is he in now?

"In a way, I've always been in the same stage—a work in progress."

We're all works in progress, and when all is said and done, that's the best way to exist. Knowing what genetics has to say about how every experience leaves an imprint on up to hundreds of genes, the process will never stop, because it's not meant to. To be alive is to join the river of evolution—the real river you cannot step into at the same place twice. The higher healing is about embracing every experience with an attitude of expanding, growing, and evolving. What keeps life going? Life itself. When we place our trust on this fact, the healing journey gets where it needs to be, an all-consuming project that expresses the joy of being alive, here and now.

PART TWO

Healing Is Now

A 7-DAY ACTION PLAN

We began this book by saying that expanding the definition of immunity was an urgent need, because every person's health is being challenged as never before. You must make sure that your immunity doesn't reach the tipping point at which stress, lifestyle disorders, and aging get the upper hand. Now you have the knowledge to follow a new model—the healing self—that will boost immunity and protect your health for a lifetime.

But knowledge is useless until it gets activated. That's almost too obvious to state. Motivating people to act tends to face a huge obstacle. Good intentions fade, and best-laid plans go astray. So we had to ask ourselves how an action plan can last a lifetime. Nothing less will produce the benefits we've been holding out as a real possibility.

The answer came to us by watching young children. Childhood development, as every parent knows, is fascinating to watch. A four-year-old is playing with paper dolls and alphabet blocks, then you turn your back for what seems like an instant, and the same child is reading books and playing hopscotch. Major changes have taken place in brain development to coordinate everything needed to learn how to read, even to do something as simple as hop on one foot with perfect balance.

Nature has arranged every step of childhood development to be so effortless that a child doesn't even know that an earlier self has been discarded for a later one—and this gave us our clue. Adopting the healing self needs to be so effortless that in a week, month, or year, major changes have occurred that feel so natural, you can't remember living any other way.

This is the philosophy behind the seven-day plan presented here in Part Two. Each day focuses on a theme that occupies your attention for that day. Monday, for example, contains recommendations about changing to an anti-inflammatory diet. There are several recommendations in the "Do" category and several in the "Undo" category—we prefer "Undo" to "Don't" because changing your lifestyle usually involves old choices you need to abandon. No recommendation is better than the others; choose whatever appeals to you.

On Tuesday you'll move on to a new theme, stress reduction, putting your focus there. If you don't want to continue with the change you made on Monday, that's fine.

After you finish the week and turn the page to the next week, the same themes will be repeated. Once more you casually choose the changes you want to make. We feel that by doing this, putting absolutely no pressure on yourself, your bodymind will enjoy each change and retain the ones that feel good. To counter inflammation, for example, one person might add nuts to her diet while another increases the amount of fiber. We can't predict which change will stick, but if these two people persist, it's inevitable that some choices will become part of their lifestyle—it's only a matter of time.

Here's the weekly schedule, covering themes you learned about in Part One.

Monday: Anti-inflammation Diet

Tuesday: Stress Reduction

Wednesday: Anti-aging

Thursday: Stand, Walk, Rest, Sleep

Friday: Core Beliefs

Saturday: Non-struggle

Sunday: Evolution

Your only obligation is to follow your own desires, selecting something to do or undo from the list of choices. We recommend that you read the entire section for that day at least once and go back to it as often as possible for reinforcement.

How will your choice turn out? Keep an open mind. This is an experiment with you playing the role of both scientist and lab rat!

For some themes, like following an anti-inflammatory diet, it will be easy to make simple changes and stick with them, hopefully on a permanent basis. In other cases, such as taking a half-hour walk every evening, fitting this choice into your schedule over a long time span may be more challenging. Just move at your own pace and always remember that your choices should be enjoyable.

Anti-inflammation Diet

Today's recommendations—choose only one.

DO

Add some anti-inflammatory foods to your diet.

Include more organic food to your grocery shopping.

Increase the fiber in your diet.

Take a probiotic supplement (see page 182).

Switch to olive or safflower oil.

Drink coffee 1 to 5 times a day, preferably at the heavier end.

UNDO

Cut down sharply on your sugar intake.

Cut out junk food and fast food.

Throw out stale food, including stale cooking oils
and leftovers more than a day old.

Reduce overall fat intake.

Reduce salt intake.

Use no alcohol.

Monday's action plan is about eating to reduce inflammation. We're targeting diet for two reasons: first, the changes you make can be incremental, which makes it easier to adopt an anti-inflammatory regimen you will stick to over time; second, the American appetite for excessive sugar, salt, fat, and processed food is seen as a major contributor to inflammation. So under "Do" we want you to add more foods that will support the healing response, while under "Undo" we ask you to cut back on the parts of your diet that aren't supporting healing.

Diet alone isn't enough to keep chronic low-level inflammation at bay. As medical science keeps discovering more and more ways that inflammation affects a host of bodily processes, it is beginning to dawn that this is a whole-system enemy that may be encroaching everywhere. The most general explanation for what creates chronic inflammation is twofold. In the first case, white cells and other immune cells swarm to fight a threat that doesn't actually need the inflammatory response. In this case the cells, having no real assignment, may start attacking the body's own cells. In the second case, there is a low-grade threat that's real but undetected by the person or their doctor. Then the immune response keeps being triggered without resolving the underlying problem.

It's basically the second case that you can help by changing your diet, which in turn can affect your intestinal tract and the process of digestion. To properly digest a meal requires an array of micro-organisms, bacteria that break down specific nutrients. Over time, this colony of bacteria has evolved into its own ecosystem within the body, known as the microbiome. We spent a great deal of time in our previous book *Super Genes* discussing the microbiome, which when you factor in bacterial DNA, amounts to an estimated 2 million genes. Compare that to the 20,000 genes you were born with and it's safe to say that we're a bacterial organism.

The importance of the microbiome, which exists primarily in the intestines but has other sites like the skin, vagina, and armpits, is immense, and it is directly affected by what you eat. These bacteria aren't invaders. The microbiome is as much your DNA as the DNA inside a heart or brain cell—in fact, human DNA is now known to contain large contributions of microbial DNA that became assimilated over eons of life on Earth.

Feel free to skip to the list of "Dos" and "Undos" for today, but some fascinating information about the microbiome has been emerging that we'd like to share. The body is open to the outside environment with every breath, and the standard medical model held for many years that the nose and sinus cavities, as the first location where microbes arrive when they are inhaled, are vulnerable sites. It's true that dust, allergens, and micro-organisms are filtered through your nose and sinuses, but no one suspected that these warm, moist mini-environments are actually alive and thriving with their own microbiome.

Yet that's the reality, and it now seems that human beings have been relating to the DNA of the microbes inside the nasal-sinus cavity in very complex ways. Actually there are two relationships going on and constantly shifting. One consists of microbe colonies interacting with one another; the other is human interaction, spanning a time as brief as a day or as long as our species has existed. People who go around with a stuffy nose or congested sinuses all the time (chronic rhinosinusitis) may not be reacting simply to something in the air, an allergen or pathogen. Instead, some kind of imbalance in this tiny microbiome may be to blame. It is surmised that bacterial activity leads to chronic inflammation of sinus tissues (a conclusion you won't be surprised to hear).

Another example is the oral microbiome that inhabits our mouth. Hundreds of species of viruses, bacteria, and fungi are involved—don't apologize if the image makes you a little queasy. These bond together into a biofilm that coats all the mucous membranes inside your mouth. Brushing your teeth then using mouthwash doesn't remove this very stubborn film, and you wouldn't want it to. This miniature ecology has evolved over the past 2 million years to keep the human species healthy, although how exactly this cooperative relationship works is by no means understood.

One theory holds that the bad bacteria (pathogens) are always present in the oral microbiome but are so outnumbered by the good bacteria that they are kept in check. Disease breaks out if the balance is reversed and the pathogens begin to overpopulate the ecology. This could be triggered by inflammation, but no one is certain. Other triggers might be responsible. To develop a reliable understanding of all microbiome locations, large and small, the Earth Microbiome Project and similar endeavors are producing a catalog of the genomes of thousands of species of microbes inside us. Back in 1972, it was estimated that bacterial cells outnumbered human cells 10 to 1, but we now know that micro-organisms are estimated to be at about a 1 to 1 parity with the body's actual cells, so mapping all their total DNA is one of the vastest projects in the history of biology.

Without repeating in detail what we covered in *Super Genes*, here are the main points that pertain to today's action plan:

- The gut microbiome is different from culture to culture. In each of us it is constantly shifting in response not just to diet, but to stress and even emotions.

- Because of its complexity and the enormous variability from one person to the next, a "normal" gut microbiome hasn't been defined yet.

- It is generally believed, however, that a flourishing, healthy gut microbiome is founded on a wide range of natural foods rich in fruits, vegetables, and fiber.

- The modern Western diet, which is low in fiber but high in sugar, salt, fat, and processed food, may be seriously degrading the gut microbiome. Other culprits include emulsifiers and artificial sweeteners.

- When the gut microbiome is damaged or degraded, bacteria begin to release so-called endotoxins—the by-products of microbial action. If these toxins leak through the intestinal wall into the bloodstream, markers for inflammation are triggered and persist until the toxins are no longer present.

From these bullet points a vast amount of information can be extracted, because wherever the bloodstream extends, which is everywhere, inflammation triggered by the microbiome can start to create problems. Today, however, we are concerned with getting your gut microbiome back into a healthy state.

The "Do" Side

On some days we will give choices that don't apply to everyone's lifestyle, but adopting an anti-inflammatory diet applies essentially to all of us. We'll reference the dietary information that we researched for *Super Genes*. The essence is to adopt a natural, organic, whole-foods diet as much as possible. With box-store outlets now offering organic produce, going organic isn't as expensive as it once was. But we are aware of the impact on the household budget of buying whole foods when processed, and fast foods are so much cheaper per calorie. But keep a few things in mind:

- *You probably don't need the calories you think you need.*

  People tend to lead sedentary or semi-active lives more and more, particularly as they age. Such a lifestyle requires drastically fewer calories than you might think. The older guidelines put a lower limit per day of eating around 10 calories per pound if you are inactive (i.e., if you weigh 150 lbs., you should consume 1,500 calories a day). An average adult who is semi-active was thought to need somewhere in the vicinity of 2,000 to 2,500 calories a day.

  But some reports on extremely sedentary lifestyles cut this figure drastically. What used to be considered a fasting diet, which began in the 1,200- to 1,500-calorie range, may well be a normal requirement for people who spend hours a day working at the computer or playing video games.

- *Cheap calories are not the same as nutritious calories.*

  America is addicted to empty calories, which also happen to be the least expensive. Sugar in the form of corn syrup and various fats like corn oil, which are extremely cheap for use in processed foods, also have inflammatory properties. The calorie curve rises in processed, junk, and fast foods while the nutrition curve—fiber, vitamins, and minerals—declines.

- *Whole foods are nature's way.*

  The debate about the unhealthy American diet is over as far as science is concerned, but people have a long way to catch up. The bottom line—no matter what foods you choose—is that the human intestinal tract, including its teeming microbiome, is adaptive to more foods than that of any other creature—we are the ultimate omnivores. This incredible adaptive ability evolved over tens of thousands of years entirely on natural whole foods.

  The spike in sugar, salt, and fat that occurred in the American diet essentially after World War II has happened too fast for our bodies to evolve and adapt. The shock of the new diet is still with us, and the resulting damage tends to challenge or even overwhelm our adaptive capacity. Hormonal imbalance, obesity, type 2 diabetes, insulin re-sistance, and excessive insulin production (hyperinsulinemia), and in-creased food allergies, including suspected gluten allergies, were all once rare and now have become endemic to modern Western society. Ignoring nature's way has come at a high price.

- *Whole foods are nonaddictive.*

It's undeniably more expensive to buy whole and organic foods, but they are satisfying to eat and do not lead to the addictive effects of processed, junk, and fast foods. Addictiveness is built into bad foods by habituation, or by developing a constant craving for high levels of sugar and salt, along with the tastes that engender cravings: sweet, sour, and salty. Every "happy meal," under whatever name, leans heavily on those three tastes.

When you switch to whole organic foods, the money spent on snacks, sodas, ice cream, and chocolate declines, helping to balance the budget. Per calorie, those foods can be among the most expensive, especially if you favor deluxe ice cream and chocolates.

A natural whole-foods diet takes care of a wide range of anti-inflammatory issues, but what about specific foods? Anti-inflammatory foods have come into favor with increasing public interest and research studies. If you are primarily interested in seeing a list of specific anti-inflammatory foods, the following are included to reinforce your knowl-edge, not to tell you that only these "right" foods belong in your diet.

Foods That Fight Inflammation

Fatty cold-water fish (such as salmon, tuna, mackerel, herring)

Berries

Tree nuts (such as walnuts, almonds, and hazelnuts, but excluding peanuts, a groundnut)

Seeds

Whole grains

Dark leafy greens

Soy (including soy milk and tofu)

Tempeh

Mycoprotein (from mushrooms and other fungi)

Low-fat dairy products

Peppers (e.g., bell peppers, various chilies—the hot taste isn't an indication of inflammatory effects in the body)

Tomatoes

Beets

Tart cherries

Ginger and turmeric

Garlic

Olive oil

In their online health publications, Harvard Medical School adds a few other items to the list:

Cocoa and dark chocolate

Basil and many other herbs

Black pepper

Other listings add the following:

Cruciferous vegetables (cabbage, bok choy, broccoli, cauliflower)

Avocados

Hot sauce

Curry powder

Carrots

Organic turkey breast (substitute for red meats)

Turnips

Zucchini

Cucumbers

Leaving aside their anti-inflammatory effects, these are all healthy whole foods, and making them a mainstay of your diet can only be beneficial. However, the science is still out on whether all of these foods actually have an anti-inflammatory effect in the body, and also what effect, if any, they have on the microbiome. Still, the fact that your genome and your microbiome respond to everyday experience strongly suggests that what you eat has whole-system consequences.

The coffee connection

Many studies have validated the health benefits of drinking coffee, and quite often the mechanism is unknown. A 2015 study of more than 200,000 subjects whose health was assessed for thirty years found a 15 percent lower risk of mortality among those who drank 1 to 5 cups of coffee a day. We've used this as our guide, but drinking at the higher end (4 cups or more a day) seems to increase the benefits. There are improvements in the risks for type 2 diabetes (perhaps connected with coffee's ability to lower blood sugar), heart attacks and strokes (probably linked to an anti-inflammatory effect), liver cancer (cause unknown), suicide (cause unknown), and disorders as diverse as gallstones and Parkinson's.

Because the most probable link with living longer is reduced inflammation, we've chosen this as the best reason to drink coffee. For longevity, it doesn't seem to matter whether the coffee is caffeinated or decaffeinated. Since coffee drinkers are more likely to smoke, it's important to note that increased longevity applied after removing smoking from the equation. (Also, tea and especially green tea are claimed to have broad health benefits, most likely due to an anti-inflammation connection, but research is less extensive and conclusive than for coffee.) Rather than jump on coffee as a magic elixir, you should add it to the list of beneficial foods while keeping the broader picture in mind.

Prebiotics

In tandem with the mounting research on the microbiome, there is skyrocketing interest in foods that keep the microbiome healthy. You've probably heard about *probiotics*, which are either foods or supplements that add beneficial microbes to the intestinal tract. On the other hand, *prebiotics* are foods or supplements containing plant fiber that nourish the microbes that already exist inside your digestive system.

As a general rule, and keeping up with the best research, you should focus first on prebiotics in order to keep the microbiome from releasing endotoxins that trigger the inflammation response. Throwing new bacteria into the mix won't be helpful if your diet is low in fiber, as the typical American diet tends to be. The government recommendation for fiber intake is 24 grams a day of both soluble and insoluble fiber, which is roughly twice what a typical American diet contains. You don't necessarily need to start counting grams of fiber, although they are now listed on the nutrition labels of processed foods.

Once you consume whole foods, especially fruits and vegetables, your fiber intake will be healthy. The most basic soluble fiber is cellulose, the undigested bulk in all plant foods. But cellulose is what the microbes in your digestive tract thrive on. Fiber has been touted for decades as a preventive for heart disease, going back to initial findings from African tribes who consumed enormous amounts of fiber and exhibited almost no heart disease. But the panacea soon faded, because other preventive factors were at work, such as abundant exercise and low stress in an indigenous tribal life compared with lifestyles in Western cultures. What makes fiber so attractive still is its broad-spectrum benefits. Not only does fiber fight against inflammation, but it also buffers the digestion of sugars (helpful in some type 2 diabetes); makes you feel full (which helps with overeating); and preserves the health of the lining of the digestive tract (which may be a crucial factor in some colorectal cancers, for example).

Eating a variety of soluble and insoluble fiber is a good idea, selecting from the following easy-to-find sources.

Soluble Fiber

Beans and peas

Whole grains (including oatmeal, whole wheat bread, and multigrain breads)

All fruits, but especially those with more soluble fiber than insoluble: apricots, grapefruit, mangoes, and oranges

All vegetables—especially the cruciferous (cabbage) family, which are very high in soluble fiber: cabbage, brussels sprouts, broccoli, bok choy, and so forth

Flaxseeds

Psyllium, a vegetable extract that is the basis of most commercial fiber supplements—also the only fiber supplement known to reduce levels of the "bad" LDL cholesterol

Insoluble Fiber

Oat bran, often taken as a supplement

Bran-based breakfast cereals

Shredded wheat breakfast cereals

Nuts and seeds

Beans and lentils

Fruits and vegetables in general

Probiotics

Probiotic foods contain living bacteria. Active yogurt is the most popular probiotic advertised on TV and sold in the supermarket, but there's also pickles, sauerkraut, kimchi (a traditional Korean fermented cabbage dish), and kefir (a fermented milk drink that tastes similar to yogurt). Including one of these foods during a meal helps to reset your microbiome by introducing beneficial bacteria that will colonize the walls of the intestine and hopefully reduce or drive out harmful bacteria. Because of the complexity of the microbiome and the huge differences from one person to another, there is no completely reliable prediction on the effects of probiotic foods. The best thing is to try them—all are completely harmless—and then look for results.

Probiotic supplements are a booming business that's expected to rise dramatically in the future. Health food stores offer a bewildering variety of these supplements, some in pill form to be taken on a full stomach, others in perishable form that must be refrigerated. There is no expert medical advice about the best probiotic supplements, for the simple fact

that the microbiome is too complex to fully understand as of now. It should also be noted that a reliable supplement that contains 1 billion bacteria will enter a gut ecology of 100 trillion microbes. Outnumbered 100,000 to 1, the supplement may have negligible impact. To be optimistic, on the other hand, any opportunity to raise the microbiome to a state of natural balance is worth taking. A supplement can't substitute in any significant way for getting your probiotics through food, yet it's an easy choice to take a supplement.

As a side note, you can augment the anti-inflammatory effect by adding a baby aspirin, or half an adult aspirin, to your daily routine. The aspirin is a proven way to lower the risk of heart attack and some kinds of cancer, such as colorectal, melanoma, ovarian, and pancreatic cancer. To date, however, the strongest evidence is confined to colorectal cancer; other results have been hit-or-miss, according to Harvard Medical School. (Be sure to consult your doctor before combining aspirin with other drugs, particularly those that have anti-inflammatory or blood-thinning properties.)

The "Undo" Side

The choices we've listed will come as no surprise to anyone who has paid attention to years of warnings about the imbalances in the typical American diet. If you can begin to eliminate the excessive salt, sugar, and fat in your diet, that's the best way to complement adding natural organic foods. But there are a few bullet points to consider:

- *Start improving as early as you can.*

  Dietary cravings get worse the longer they persist. Children who start out life on a high-sugar, high-salt diet soon adapt to it as their default normal diet. You may not still be young, but as a parent you need to set a good example for the whole family.

- *Don't let age catch up with you.*

  As people age, their diet typically declines. They move in the direction of convenience foods, although these aren't necessarily bad now that the freezer section includes lots of healthy choices that are much lower in sodium and fat than even a decade ago. There is also

a tendency for older people to favor a mono diet, or close to it, in which only a handful of foods are preferred. This is very unhealthy as you age.

The intestinal tract becomes less efficient as we get older, which means among other things that we don't assimilate vitamins and minerals the way we did when we were younger. In some studies the effects of dementia and memory loss have been reversed dramatically by restoring trace minerals like manganese and zinc to the diet. Even doctors rarely think about mineral deficiencies, but if you are older, it's good advice to take a multipurpose vitamin that fulfills the daily requirements for minerals. Even better is to keep eating a well-established natural whole-foods diet.

At the same time, when age-related decline in kidney function is added to the picture, there can be a lack of water-soluble vitamins (vitamin C and the B complex), which get passed out in the urine. Supplements of these vitamins are helpful, particularly if your diet hasn't gotten rid of things that need undoing.

- *Consider no alcohol.*

    Alcohol has a fixed place in American social culture, and most people use it one way or another. The benefits of alcohol in preventing heart disease seem to be sound, as long as intake is limited to one drink a day—usually a glass of wine at dinner. Research now suggests that the fabled benefits of red wine in the French diet aren't unique—it's the alcohol itself that is beneficial.

    Harvard Medical School's website states that moderate alcohol consumption is even anti-inflammatory, which seems counterintuitive. The red noses of heavy drinkers is a sign of both inflammation and liver damage. The anti-inflammatory benefits of alcohol vanish and turn into inflammation as the result of excessive consumption.

    For many people, one drink can easily lead to two or three. In addition, a certain percentage of drinkers will become alcoholics; moreover, as people age, loneliness, boredom, and a sedentary lifestyle can lead to more drinking. So in the overall picture, there are enough pitfalls to alcohol that you should seriously consider them. We'd be happier if alcohol was cut to a bare minimum (a glass of wine when eating out in a restaurant).

- *Keep it fresh.*

One of the reasons that antioxidants have become popular is to counter roaming oxygen in the bloodstream known as free radicals—in other words, oxygen atoms that quickly latch on to other chemicals. This chemical reaction is totally necessary in the healing response at wound sites, for example, so it is too simplistic to claim that free radicals are "bad."

However, a simple way to outmaneuver the whole issue is to eat fresh food and throw out stale cooking oils, leftovers more than a day old, freezer-burned food, and the like. Staleness is associated with oxidation and also a host of micro-organisms that potentially have an inflammatory effect. In any case, staleness isn't what you want in the first place. Since cold-pressed virgin olive oil is especially good for anti-inflammation but also is one of the oils that goes stale quickly upon exposure to air, it's best to store the bottle in the refrigerator and keep at room temperature only the amount you need for two or three days.

Just as it's not enough to focus totally on a laundry list of anti-inflammatory foods, don't fixate on other foods as somehow bad or un-healthy. We want you to use common sense when reading the following list of foods that have been tagged as having inflammatory properties.

Foods to Limit or Avoid

Red meat

Saturated and trans fats (e.g., animal fats and the hydrogenated vegetable fats found in many processed foods)

White bread

White rice

French fries

Sugary sodas

To these, other reliable sources add the following:

White sugar and corn syrup (frequently hidden in processed foods that aren't primarily sweet)

Omega-6 fatty acids (see below)

Monosodium glutamate (MSG)

Gluten (see page 125)

Our feeling is that an anti-inflammatory diet has to be better than an inflammatory one, because the foods that are proven risks—junk food, fast food, fatty and sugary foods—also lead to inflammation. The link between inflammation and chronic disease is too strong to ignore, and paying attention has many benefits.

A note on omega-3 and omega-6 fatty acids: For decades the public has been conditioned to regard cholesterol as a "bad" fat even though cholesterol is a biochemical found in every cell and is necessary for cellular development. The same has happened in reverse with omega-3 fatty acids. We go along with the general recommendation that cold-water fish high in omega-3s, like salmon and tuna, are beneficial. But the story is more complicated than this.

There is another group of fatty acids known as omega-6. Both omega-3s and omega-6s are needed in the diet; our bodies don't produce them. It turns out, however, that omega-6s in excess have a strong link to inflammation. Moreover, since both groups often occur together, the harmful effect of the omega-6s can undo the benefits of omega-3s. In short, the two must be kept in balance. All Western diets are too high in omega-6s because of the heavy use of polyunsaturated cooking oils. Yet these oils, made from vegetable sources—corn, soy, sunflower, and so on—were once considered the healthiest ones, with lowering risk factors for heart attack as the primary support for this claim.

Today the evidence has strongly moved in another direction. Studies of indigenous peoples (who use few processed vegetable oils and eat no processed packaged foods) indicate that the ratio of omega-6s to omega-3s in their diet is about 4:1. In contrast, Western diets are 15 to 40 times too high in omega-6 foods, with an average ratio of omega-6s to omega-3s of 16:1. At such high levels, the omega-6 fatty acids block the benefits of the omega-3s. Genetic studies aren't easy to come by in this area, but it's speculated that we evolved in hunter-gatherer societies to consume a diet even lower in omega-6s, with a ratio of omega-6s to omega-3s closer to 2:1. In the body, getting closer to a 1:1 ratio seems ideal, according to some experts.

Among foods high in omega-6s, cooking oil leads the way, but there are others, as follows.

Main Sources of Omega-6 Fatty Acids

Processed vegetable oils—highest are sunflower, corn, soy, and cottonseed

Processed foods using soy oil

Grain-fed beef

"Factory-raised" chicken and pork

Non-free-range eggs

Fatty cuts of conventionally raised meats

Unfortunately, the polyunsaturated oils that are a major part of standard disease prevention turn out to have a serious drawback in terms of inflammation. The only vegetable oil that is low in omega-6s and also high in omega-3s is flaxseed oil. Safflower, canola, and olive oil aren't particularly high in omega-3s but are the lowest in omega-6s among commonly sold vegetable oils, with olive oil the best.

Adding to the confusion, "bad" saturated fats like lard, butter, palm oil, and coconut oil are low in omega-6s. This is one reason why standard prevention has begun to recommend a balance of saturated and polyunsaturated fats. But the real culprit, it seems, isn't so much the food we eat in its natural state but processed foods. Soy oil is cheap and readily available, lending itself to use in hundreds of packaged foods. Beef from cattle raised in feedlots on grain to achieve maximum bulk in the shortest period of time is much higher in omega-6s than grass-fed beef (not to mention the widespread use of antibiotics and hormones in the beef and dairy industry). Also high in omega-6s are pork and chicken produced on conventional grain feed in the "factory" system, along with factory eggs.

If you are going to eat meat, we advise switching to grass-fed beef, along with naturally fed (also called pastured) chickens and their eggs.

"Free range" isn't always reliable, since the birds might still be receiving some conventional feed. Of course, this isn't always an easy or viable option. Grass-red beef and poultry can be expensive and often are only found in specialty stores. So do what you can. Overall, rebalancing the fatty acids in your diet comes down to some easy steps once you are conscious of the issue. Don't fixate on this one aspect of your diet— everything on the list is compatible with the whole foods you should be moving to, step-by-step.

How to Balance Fatty Acids

- Cook with safflower and olive oil; canola oil isn't as good but is acceptable.

- Eat unsalted or low-salt tree nuts, including walnuts, almonds, pecans, and Brazil nuts. Limit the amount of fatty nuts, such as cashews and macadamias, as well as peanuts.

- Eat seeds, including unsalted chia, sunflower, pumpkin, hemp, and flaxseeds.

- Eat fatty fish—no more than 6 ounces per week as well as mycoprotein-based products. If vegetarian, eat more tree nuts that are lower in fat, such as walnuts and almonds, and seeds.

- Avoid packaged foods with soy oil high in the list of ingredients.

- Don't cook with soy, sunflower, or corn oil.

- Cut back or eliminate conventionally raised beef, pork, and chicken.

- With any meat and poultry, buy lean cuts and trim the fat from other cuts.

The interaction of food with the bodymind is at once fascinating and complex. We wanted to give you some in-depth information, but when it comes to the practicalities, you should move at your own pace and remember that transforming your diet is a marathon, not a sprint. What's important aren't the choices you make but the good choices you stick with over time.

That's why our action plan for dietary changes involves some of the simplest, most direct steps in healing the whole system that you can take. Everyone should give them high priority. But if present trends continue, a lot more focus is going to be placed on the microbiome and its connection to inflammation. Diet is only one factor, which will come as no surprise in our whole-system approach. To really do more to heal and balance your microbiome, you need to think in terms of the bodymind as a whole. Here is a helpful list that gathers the best information to date about lifestyle and the microbiome. Such a lifestyle would include all the choices covered in today's action plan, but it means going a few steps further.

The Optimum Lifestyle for a Healthy Gut Microbiome

Eating less fat, sugar, and refined carbohydrates

Adding sufficient prebiotics on which bacteria feed: fiber from whole fruits, vegetables, and grains

Avoiding chemically processed foods

Eliminating alcohol consumption

Taking a probiotic supplement

Eating probiotic foods like yogurt, sauerkraut, and pickles

Reducing foods with inflammatory effects

Focusing on foods with anti-inflammatory effects, like freshly squeezed orange juice

Managing stress diligently

Attending to "inflamed" emotions like anger and hostility

Checking for medical causes of inflammation such as yeast infection and stress

Controlling weight gain

As you can see, being totally free of inflammation is the same as leading a healing lifestyle in general. That's why we focus today on diet as the single best way to deal with the problem. In other aspects of your life, such as losing weight or stress management, inflammation doesn't really need to be singled out—these measures are for the whole system and general well-being.

Tuesday

Stress Reduction

Today's recommendations—choose only one.

DO

Meditate.

Go to a yoga class.

Practice mindful breathing.

Schedule downtime and quiet time.

Practice being centered.

Recognize the stages of stress (page 81).

UNDO

Stop adding to a stressful situation.

Refrain from ignoring stressful events in your life.

Walk away from stress as soon as you can.

Resolve a repeated stress.

Examine a problem you have been putting up with out of frustration.

Turn irregular habits into a regular routine.

Unlike chronic inflammation, which very often lurks beneath the sur-
face and goes undetected, stress is an enemy that hides in plain sight. On

any given day, the average person knowingly confronts the same stressors over and over: excessive noise and haste; overlapping demands at home and at work; sensory overload everywhere; frustrations driving a car in traffic; and too little time in the day to accomplish everything that needs to be done. What external stressors all have in common is pressure, and everyone knows what pressure feels like. If external stressors were the real problem, resolving stress would not be much harder than removing a rock from your shoe—having felt the discomfort, you'd deal with it as soon as possible.

But as we know, stress is often much more complicated than that. The fact that we all put up with so much stress is testimony to how badly we deal with it. Today we want you to turn the corner and begin to seriously reduce the stress in your life. You may be good at putting up with daily stress, but in tiny incremental steps your cells are adversely affected anyway. On page 81 we listed the three stages of how stress affects people, first psychologically and mentally, then in their behavior, and ultimately in the form of physical damage. But waiting until the third stage, when symptoms like hypertension and digestive problems appear, would be very shortsighted. Stress will have gained the upper hand long before.

When Stress Gets the Upper Hand

If people complain about stress all the time, constantly hear about its damaging effects, and yet still do little to nothing about it, what has gone wrong? The choices we've listed for today aren't novel or surprising. Meditation and yoga are now so familiar that taking them up should be far more widespread than it is. Scheduling downtime and quiet time during the workday should be routine. Learning to stay centered in a stressful situation should be a coping mechanism learned from a young age.

Clearly the first and most important step in reducing stress is a change of attitude. Otherwise, your ability to deal with the pressures of everyday life will remain stuck where it is now, a half-hearted stab at coping without much in the way of results. In a way, the situation is akin to crash dieting. As we mentioned earlier and as most people realize, temporary diets don't keep weight off. The number of dieters who successfully keep more than five pounds off for two years is under 2 percent. But in the face of these dismal facts, Americans diet constantly, and small fortunes are made by promoters of the newest fad diet. In other words, people keep doing more of what never worked in the first place—and many people do the same with stress.

So that you can begin to change your attitude toward stress, here's a list of the things that never worked in the first place.

Why Stress Keeps Winning

INEFFECTIVE RESPONSES TO EVERYDAY STRESSORS:

We consider it normal to be a little stressed.

We feel helpless in the face of external forces.

The signs of distress (irritability, fatigue, mental dullness) are ignored.

Our coping mechanisms are too limited (see page 74).

We think that putting up with stress is harmless.

We're in denial or just plain unaware of how stressed we are.

We've heard that it's possible to thrive on stress.

These beliefs and actions are self-defeating, but each holds a grain of truth. If you live in a noisy city or work on a construction site, the racket all around you is beyond your control. Putting up with stress isn't harmless, but if you're stuck in traffic or have a newborn baby in the house, there's not much else you can do. No one thrives on stress, not at the cellular level, but some ambitious, successful people claim that they owe their success to an appetite for high-stress situations in which they can prove that they are winners. These grains of truth serve as cover for a reality that people aren't willing to face: stress is the epidemic of modern life.

For this point to really sink in, let's illustrate it. First, a typical day for A, a young husband and father rising in his career. A wakes up a bit late and rushes to get ready for work. He hears the kids fighting in another room and yells for them to stop. On his way out the door he kisses his wife and says he's in too much of a hurry to grab breakfast. Traffic is bad, so he's not in the best mood when he gets to the office, where his boss is staring at his watch and reminds A that there's a big deadline coming up.

After a staff meeting at which the whole team is pushed to produce results, A slows down enough to take a coffee-and-doughnut break. A little guiltily he loosens up with a drink at lunch, and he feels less tense the rest of the afternoon. The afternoon commute isn't so bad, and A feels pretty good when he gets home. He settles into his familiar domestic routine, spending a few minutes with the kids and a few hours online. His wife has learned to put up with this. A gets steamed when he goes to a provocative news site—damn politicians. Before going to bed he attacks some work he brought home with him. A and his wife still have an active sex life, but they're both too tired tonight. There's always the weekend.

By no means is this a parody of how millions of people spend a typical weekday. Every event is a stress point, but by society's standards, A is leading the good life, or doing what it takes to get there. A generation ago, when stress was a new topic, someone's typical day might include chain smoking, imbibing considerably more alcohol, and more burdens on women in the home. Medicine knows a vast amount about the effects of stress all the way down to the epigenetic level, where negative experiences leave marks that alter genetic activity. Yet this knowledge hasn't translated into how we lead our lives. Today we want you to ease into a self-aware approach to reducing everyday stress in your life.

The "Do" Side

All the recommended choices related to stress focus on getting out of sympathetic overdrive, which we devoted an entire earlier chapter to. The opposite of stress is relaxation. Practices like meditation and yoga go far beyond simple physical relaxation, even beyond finding mental peace and quiet. But relaxation is a start, because without it, the bodymind is dealing with disruptions due to stress, and this preoccupation blocks the ability to support subtler experiences. Both authors strongly endorse the Eastern wisdom traditions, which have higher consciousness as their foundation. We endorse the higher healing that arrives with higher consciousness. But on the basis of first things first, people need to return to a default state of relaxation throughout the bodymind.

Meditation: The one stress-reduction practice we could list for every day of the week is meditation, because of its holistic benefits. So far in this book we've been open ended about which type of meditation you might prefer. Mindfulness meditations are popular; meditating on the breath is effortless; meditating on the heart is appealing to many who have an inclination toward devotion. There are countless books and websites that allow you to explore the whole subject.

There are scant experiments that compare one meditation to another in order to efficiently prove that there is a "best" meditation. In truth, *best* isn't an applicable term. The style of meditation that you are comfortable with and that will become a lifelong practice is by definition the best meditation. People drop their meditation practice when they no longer feel any benefits; they keep with a practice if they sense continual personal growth. None of this is predictable. (Sometimes meditation gets dropped simply because life is going well, which is taken as a sign that meditation has already done its job.) In terms of the most proven benefits, the nod probably goes to mantra meditation because of its ancient roots in India, where literally hundreds of mantras have specific effects; the ultimate effect is enlightenment, or total self-awareness undisturbed by external events.

A simple mantra technique that has no religious connotations is as follows:

- Sit in a quiet, softly lit room. Close your eyes for a minute or two. If you feel sleepy, lie down and nap rather than starting to meditate.

- When you feel centered and your breathing is relaxed and regular, silently say the mantra *So hum*.

- Repeat the mantra for 5 to 20 minutes, depending on individual circumstances and how much you enjoy meditating.

- Do not repeat the mantra mechanically—this isn't silent chanting. Instead, say *So hum* when it comes to mind to repeat it. There can be gaps as short as a few seconds or as long as several minutes. Mantra meditation calms the mind not by stopping the thinking process, but by allowing the mind to settle into a naturally quiet state. There is no question of forcing and nothing mechanical. No magic is involved. The repetition settles the mind through its natural inclination to be quieter and calmer.

- It doesn't matter if thoughts intrude—they always will. Thoughts are a natural part of meditation. Simply return gently to the mantra. There are no minimum goals for how often the mantra gets repeated. If you say it once and then doze off, that's a good meditation. You needed the rest. If you say the mantra once and go into deep meditation, that's also good, and everything in between.

- Meditation allows the bodymind to rebalance itself through stress release. In the course of stress release, any sensation or thought can arise. This is normal, effective meditation. If a physical sensation is so strong you cannot think the mantra easily, then put your attention on the area in your body where you feel the sensation. Let your awareness be with the sensation without trying to change it in any way. After a moment or so, the sensation will fade. If it doesn't and you feel persistent discomfort, open your eyes for a few minutes. If the discomfort doesn't fade away, lie down and rest until it does. (Actual persistent pain requires a doctor's consultation.) Don't mind negative thoughts: they will come and go. This is a natural aspect of meditation. However, if the negative thoughts feel overpowering, open your eyes and breathe easily until the thoughts subside. Once the intense thoughts have subsided, you can resume meditation.

- After the designated time is up, relax and enjoy the meditative state, eyes closed, breathing easily. To assimilate the relaxed state even more completely, lie down for 5 minutes. Don't rush back into full activity; ease back into your daily routine if circumstances permit.

- How often you meditate is up to you. Twice a day, morning and evening, is desirable once you choose to make meditation a permanent part of your lifestyle. In support of their practice, many people join a meditation group or go on meditation retreats. Again, this is a personal choice, but one benefit is that with group support, you are less likely to let your meditation practice drop off.

Mindful breathing: Today this technique is used to counteract the feeling of being stressed. We've already mentioned it in connection with being mindful at the office (see page 41). So that you don't have to thumb back to another page, we repeat the instructions here:

- If possible, find a quiet, softly lit room where you can be alone, although this isn't a necessity.

- Close your eyes and center yourself.

- Now take deep, relaxed breaths on the count of 4 for each in-breath and 6 for each out-breath. If this becomes an exertion or you begin to gasp, don't force anything. Breathe normally until you catch your breath, then return to mindful breathing.

- Continue for a minimum of 10 breaths. If you feel the need for more, continue mindful breathing for 5 to 10 minutes.

The "Undo" Side

Today the "undo" choices are about detaching yourself from stressful situations when you might otherwise be tempted to stay. These are often minor things that cause momentary tension. But even these can cause a stress reaction that you don't need. The key is to pay attention to your feelings and the sensations you might feel in your body. Check in with yourself several times today and ask if you feel tense, uncomfortable, tight, or pressured. This feeling can be either physical or mental: as far as stress goes, they are equal. Your goal today is to extricate yourself from a negative situation, find a way to be alone, and regain a relaxed, centered state.

When stress is more than incidental, more is called for. We take seriously the fact that stress is winning in many if not most people's lives. Therefore, to undo your entangled relationship with stress, we need to discuss the problem and its solution in depth.

Stress in Depth: The Inside Story

External stressors are generally the kind that get the most attention from researchers. Mice are useful subjects in the laboratory, but they don't have an inner life comparable to that of humans, so stress experiments with mice have centered on external physical stressors. In one famous experiment, mice were placed on a metal plate that emitted tiny harmless electric shocks. The shocks were administered at random, and after only a few days, the mice showed extensive damage to their immune system. They behaved nervously and erratically; some were weak to the point of total exhaustion or death.

The reason that harmless shocks caused this drastic deterioration is an invisible factor: *unpredictability*. The anticipation of the shocks was like the sword of Damocles hanging over the animals' heads. Being unable to predict the future, while knowing that another shock was inevitable, kept the mice in a perpetual state of inner stress. As applied to humans, we've already mentioned that stress becomes worse when it is random, unpredictable, repeated, and out of a person's control. But the mouse study made another critical point: inner stress is as potent, or more potent, than external stress. The anticipation of pain distresses us as much as the pain itself.

This provides the key for reducing stress: approach it from the inside. You can't control any number of external stressors, but you can control your perception and interpretation. Imagine the difference between going to a concert and hearing the cymbals give a resounding climax in Tchaikovsky's *1812 Overture*, which you enjoy and welcome. Contrast that with a stranger walking up behind you and crashing a pair of cymbals in your ear. Same external stimulus, very different inner response. Pleasure turns into an intrusive assault.

Previously we gave the "baby solution" to acute stress (page 74) based on what parents of a newborn infant can do to bring their stress levels down. Now we want to extend those strategies to chronic everyday stress, which is the kind that causes the most damage over a long period of time. By changing your perception and interpretation of external stressors, you can greatly reduce the effects of stress.

Randomness and Unpredictability

These two factors are related, since by definition random events are unpredictable. Part of letting stress win is our appetite for shock and surprise. As much as disasters and catastrophes are terrible occurrences, the news cycle has gone from one hour of major network news a night to 24/7 cable and Internet news that reinforces a craving to watch bad news over and over. Violent video games and action/adventure movies feed the same craving in imaginary ways. But the burst of adrenaline that is triggered by the stress response doesn't know the difference between real and imaginary. At a certain point, even if you don't become an adrenaline junkie, somewhere inside you probably have a positive image of a life built upon intense conflict and action (more likely if you are a male).

Taken altogether, randomness has become the everyday chaos we have all adapted to. You need to view the chaos as a factor that increases your stress levels, not an inevitable aspect of life. Of course, life is always unpredictable, and there is such a thing as creative uncertainty. Not knowing the next painting or piece of music one will make is part of the pleasure that comes with being creative. But in daily life, bringing chaos under control is still important.

Here are some steps to consider:

- Make your daily routine more regular. Get up and go to bed at the same time every day. Eat three meals a day at regular times.

- Develop a predictable lifestyle and bring erratic behavior under control. This is very important as a parent with young children, because predictability builds trust. At work, predictability builds loyalty and cooperation. In relationships, it builds intimacy.

- Being predictable isn't the same as being boring and without originality. Instead, you want to be predictable in the following ways:
  - *You don't show anger and frustration.*
  - *You don't criticize people in public.*
  - *You are responsible.*
  - *You carry out what you promise to do.*
  - *You can be counted on for follow-through.*
  - *You welcome open communication.*
  - *Your door is always open.*
  - *You let others have their own space.*

- Having established yourself as predictable, you encourage other people, especially family members, to follow your example.

- Guard against future risks (e.g., with adequate insurance, following health prevention, maintaining your car in good working order).

- Develop a support network to help in times of trouble. Do your part to support others.

- Face crises head-on. As the situation unfolds, talk about what's going on with your family and friends. Don't isolate yourself or tough things out by going it alone.

Lack of Control

Stress is aggravated when you feel you aren't in control. In animal experiments, control always belongs to the experimenter, but in nature animals organize themselves in societies where dominance is clear cut. An alpha male in a monkey troop spends time and energy preserving his status, but what doesn't change is that a single male will have such a status and that subordinate males will find their place in the pack and accept it. In humans the situation is so complex, however, that animal models often seem irrelevant. The mythical mailroom boy dreams of rising to become CEO—unlike animals, we wish, hope, aspire, and strategize.

Control, then, is about matching our inner conception with what's going on around us. If you feel in control on the inside, you *are* in control. External events may not put you in a leadership role, but that doesn't matter compared with the ability to cope with stress by not losing control. Imagine a hundred cars stuck in a bad traffic jam. If one could hook up each driver to monitors for heart rate, blood pressure, brain activity, and respiration, there would be a hundred different responses, each dependent on an inner interpretation of the event.

At the negative end of the spectrum the most stressed-out drivers would be responding in any of the following ways:

- They resent being inconvenienced.

- They dwell on how often they get trapped in traffic.

- They expect things to work out the way they want them to, and when they don't, frustration boils over.

- They give in to reflex anger.

- They blame other drivers, who are idiots.

- They become sulky and irritable with other passengers in the car.

- They get anxious about being late.

It's normal to feel some of these things, but they rise in intensity with people who tend to be type A personalities. You don't have to be a so-called control freak to be stressed when a situation is beyond your control. However, if you demand of yourself that you must always be in charge, you will be at a disadvantage when coping with situations that don't meet your expectations.

Controlling personalities are difficult to live with, all the more because they frequently think that their way is the only way—in fact, this is a hallmark of the controlling personality. Another hallmark is that they always find a way to blame others while excusing themselves. They are perfectionists about detail and will be just as critical over a misspelled word in a report as over a failed project. Their demands are never totally met; they give praise grudgingly or not at all; they expect others to live by the same values and standards they set for themselves (as in the boss who says, "I don't ask you to do anything that I wouldn't do myself"). Emotionally, they are tightly wound, anxious, and afraid to show emotion because it reveals weakness and vulnerability.

This description offers a general warning about the responses that do not work when you feel that a situation is going out of control. To some degree, all of us are tempted to impose our will, to make demands on others, to insist that our way is the only way, and so on. Seen from the inside, however, the root cause is anxiety and fear. To bring anxiety to rest, it is certainly necessary to regain control, with inner control coming first, then making an effort to bring the outside situation back from the brink of disorder.

Here are some steps to consider:

- *Learn to center yourself.* This is a skill that develops naturally with the practice of meditation. But everyone has from time to time experienced feelings of being centered—that is, calm, quiet, alert, observant, and grounded. For many people the feeling is centered in their chest.

- *Learn to recognize when you are not centered.* This state is also familiar to everyone. It is marked by anxiety; racing thoughts; uncertainty; feeling rattled by the outside situation; a pounding heart; shallow, ragged breathing; butterflies in the stomach; and muscle tightness and tension.

- *Develop the ability to get back to your center whenever you are thrown out of it.* This ability follows from the first two points—once you recognize that you aren't centered, you can return to the feeling of being centered again. To do this, several simple techniques are useful:
  - *Recognize the stressor.*
  - *Walk away from the stressful situation.*
  - *Find a quiet place to be alone.*
  - *Close your eyes and place your attention in the region of your heart.*
  - *Use mindful breathing: Take deep, regular breaths, counting 4 on the in-breath and 6 on the out-breath.*
  - *If you have the time, meditate after you feel calmer and more centered.*
  - *Continue the above until you are back in your comfort zone.*
  - *Don't rush back into the stressful situation. Give yourself a few hours, or preferably a day, to remain in an unstressed state.*

- *If you find yourself in a situation at work in which you have no control, do something about it.* Companies are beginning to realize that workers thrive when given freedom of choice to make their own decisions and accept more responsibility. It's not mandatory that you remain in a job where higher authority controls everything down to the smallest detail and a strict code of rules is enforced. Try asking for more decision-making power and freedom to offer your own solutions. If these requests are rejected, look squarely at where you stand and make plans accordingly.

- *Examine your own controlling behavior.* Take an honest look in the mirror and move toward being more accepting and nonjudgmental, less critical and demanding. These are the most glaring traits of a rigid approach to self-control.

- *Pay attention to relaxing* and being less demanding on yourself.

- *Learn to bend with the situation* before you step in to interfere and bend things to your will.

- *Find ways to be playful.*

- *Put a high value on making someone else happy.*

Repetition

Stress is cumulative; the more often it's repeated, the worse the damage it causes. A straw wouldn't break the camel's back if thousands of straws hadn't preceded it. This lesson is so simple and self-evident that one would think it didn't need to be learned over and over. But subjecting yourself to repeated stress is likely to be something you do without thinking about it. Old married couples have the same arguments for years or decades until it turns into a ritual. Politicians raise our blood pressure by lying and sidestepping the issues, as if this hasn't been true since politics began. Parents raise their voice at misbehaving children, who ignore them or stop misbehaving for a little while until they misbehave again.

Self-inflicted stress is usually marked by repetition. It falls under the category, already touched upon, of disregarding futile behavior, or "doing more of what didn't work in the first place." In the same vein, we continue to put up with things that stressed us out in the first place. It's the passive side of the syndrome: the wife who sighs when her husband puts her down for the thousandth time; the mother who can't keep the kids from fighting; the office worker who grits his teeth under an abusive boss; the disruptive student who has turned getting sent to detention into a habit.

The passive side of the syndrome is victimization, allowing bad things to repeat themselves because you feel you deserve it or can't make it stop. The active side of the syndrome is willfulness, stubbornly repeating the same self-defeating behavior because you insist on getting things to turn out the way you want them to. At the cellular level the story is the same from either viewpoint. A low-grade stress response keeps returning over and over.

Having seen what doesn't work, what does? We feel, and have advised in previous books, that you need to assess what you can fix, what you must put up with, and what you should walk away from. Most people endure repeated stress because they can't make up their minds. They waver between these three alternatives, sometimes making a stab at fixing things, other times putting up with a bad situation (the most common default response), and only walking away if worse comes to worst. Domestic abuse is a notorious example of this confusion, and even when the abused spouse manages to walk away, this often turns out to be only a temporary respite before they return. Short of such an extreme, however, all of us tend to endure situations of repeated stress because of indecisiveness. The kind of stress that keeps getting repeated may start out as something minor, but a steady drip-drip leads to a buildup, and then it's not the stressor itself that becomes the main problem, but the suppressed anger, resentment, and frustration that take their toll.

Indecision keeps you in suspense, which is the same as anticipating pain, which, as we've alluded to, has been shown to be as stressful as actually experiencing pain. Decisiveness, on the other hand, restores one's sense of being in control. There is no guarantee that the outcome will be entirely good, but instead of waiting and anticipating, you can get on with life. Here are the criteria we recommend using when you face a mounting stress that keeps being repeated:

- *Finding a fix*

    The first and best choice is always to seek an improvement. Some repeated stresses are external, such as trying to work in a noisy, chaotic office or commuting in heavy morning traffic. But the vast majority of repetitive stresses are human, and of these, most occur in relationships. So what do you do to improve a relationship that has hit a rocky place or a work situation in which someone you cannot avoid is creating constant stress?

    Step 1: Assess the chances in favor of a successful solution. The key question is whether the person at the other end of the problem is willing to listen, wants to change, can negotiate reasonably without getting angry and resistant, and can be trusted to keep faith with the bargain the two of you have reached. This is asking a lot, and you need to turn the tables and ask the same questions of yourself. You are responsible for your own emotions and responses. Blame comes from an emotional level that will always wreck any negotiation. Guilt will lead to appeasement, which ultimately leads to resentment. Also part of your assessment should be how rigid the impasse is. If you have reached the point of noncommunication or worse, freezing each other out, there is no solution in sight. You need to restore some level of communication first before facing any other alternative.

Step 2: Write down the pros and cons of each possible solution. Take your time revising and adding to your lists. A fix-it that truly works needs long consideration. Be as rational and objective as possible. A good angle is to pretend that it's not you who has the problem but a friend who has asked for your advice. What would you tell your friend, pro and con, about the solutions that seem possible? As you make out your list, check to be sure that you would be sharing the burden equally after adopting a change.

Step 3: Present the solution that came out on top in your deliberation. Don't bring the list to the table, and don't offer multiple possibilities—that simply leads to confusion. Even though you have personal issues at stake, don't let this first engagement turn into a gripe session. There's always a temptation to itemize all the things that have gone wrong back from day one. Resist this temptation. Most of the time, the other person already knows that a problem exists. Yet often the words "We need to talk" will come as a shock to the other person. It's generally best to limit the first engagement to no more than 15 minutes. You want to reach a specific goal—the solution you have in mind—and the other person deserves time to absorb what is happening. The instigator of change always bears the responsibility for heading the negotiation, which means keeping a cool head and being as fair as possible about the other person's viewpoint. Finally, if you are the one who is going to initiate the fix, wait until a calm moment when various issues aren't flaring up. The very worst time to bring up problems is when you are arguing, criticizing, under the influence of alcohol, or feeling blame and guilt.

Step 4: Having reached an agreement, follow through with your half of the bargain while asking the other person to do the same. Negotiations never succeed unless both sides feel they have won something, come away feeling safe and secure, and find a way to hold on to their dignity. Win-win isn't just an ideal; it's the only acceptable outcome, because in win-lose, the losing party will always act badly, given enough time. Remember, you are

responsible only for your side of the solution. It's not your place
to nag or remind the other person, to monitor their compliance,
or to attach blame if the solution isn't working. Backsliding
is part of everyone's tendency to resist change. The best tactic
is to pencil in a follow-up meeting right when the solution is
agreed upon. This way you eliminate the tension of watching
and waiting to see if the other person lives up to their side of
the agreement. Finally, be honest with yourself if the solution
isn't working. Instead of giving up, renegotiate; this time ask
the other person what their best solution is. Compromises are
easier to reach once two people get to the stage of "We tried it
my way, we tried it your way—now what?"

- *Putting up with a bad situation*

Most problems get worse when they are allowed to fester, and yet
we all tend to put up with bad situations out of passivity, inertia, or
aversion to conflict. The bad situation *is* the conflict. Keeping quiet
about it or going into denial only pushes the conflict underground.
Because we tend to wait too long, problems erupt into open hostility,
and then negotiations become much harder. The reason that couples
fail to reconcile their differences usually isn't because those differences are drastic but because the time for easy answers has passed. If
you feel today that you are putting up with stress in your relationship
or at work, you are already past due to seek a solution.

Yet there are times when your best solution is to stick it out. Having explored the possibility of finding a fix, but without success, you
need to sit down again with pencil and paper to make a list of the
pros and cons of putting up with the situation. Often there are outside factors—a frustrated spouse may have to think of the children,
a disgruntled employee may see no other job prospects in sight. Nobody is completely free and unencumbered. You might want to use
the headings "Good for me," "Good for us," "Bad for me," and "Bad
for us" in your deliberations. Emotionally, most people look upon
putting up with a stressful problem as a loss or defeat, victimization,
or martyrdom. It's hard to escape these feelings, which have some
basis in reality, since you did fail to reach a solution.

You must focus on the positive side of staying and sticking it out.

Spouses find ways to live together in less-than-happy circumstances, and one key is knowing that this is their decision, not a trap they've fallen into against their will. In your deliberations you want to reach the stage where you are content with your decision, as much as that is possible. The columns devoted to "Good for me" and "Good for us" need to have legitimate entries, not excuses. Putting up with a bad situation is always a compromise. What you give away will feel much worse, however, if you aren't firm in your decision. It's like the difference between giving ten dollars to the homeless and having somebody steal ten dollars from you.

Finally, ask yourself if you are using any of the following bad reasons for staying:

~ *I have no other choice.*

~ *I'm afraid to leave.*

~ *I can't take care of myself alone.*

~ *I'm suffering, but that doesn't matter.*

~ *I have to be loyal, no matter what.*

~ *This is all my fault.*

~ *Just give it more time.*

These self-defeating responses are born of fear and guilt. When any come to mind, step back and rationally ask, "Is this really true?" Remember, your goal is to make a decision in which putting up with a bad situation is as positive for you as it can be under the circumstances.

• *Walking away*

The third option is to make a decisive break. As with the decision to put up with a bad situation, walking away generally comes too late—it is emotionally forced on you when you have reached your limit to cope. We're not being judgmental—there can be many good reasons for walking away, the best one being that you have decided to stand up for yourself. What is needed, as always, is a decision you feel good about, not a last resort or an act of desperation.

Take out your pencil and paper and list the pros and cons of leaving. It's helpful to add a third column headed "What will happen next?"—the consequences of walking out on a relationship or quitting a job cannot be minimized. Ruptures always create wounds; deep wounds always take more time to heal than you anticipate. The positive side of walking out sometimes leads to a honeymoon period in which you are relieved to be free of tension, discord, hostility, and general stress. Just as often, however, the honeymoon leads to a backlash emotionally, accompanied by depression, guilt, and anxiety.

We aren't foreseeing doom—you just need to be armed with psychologically realistic expectations. The backlash of walking out varies with each person. Unfortunately, it seems to be human nature that walking out brings up selfish motives. Looking out for number one, generally tinged with revenge if a marriage is dissolving, becomes a driving motive. Try not to fall into the trap of self-preservation at all costs. There are strong elements of fear and insecurity at work here. Be aware of what's really going on inside you, because as long as anger and revenge are relied on as a driving motive, they mask the hurt that needs healing.

Anti-aging

Today's recommendations—choose only one.

DO

Meditate.

Join a social support group.

Strengthen emotional bonds with family and close friends.

Take a multivitamin and mineral supplement (if you are age sixty-five or older).

Maintain a balance of rest and activity.

Explore a new interest.

Take up a challenging mental activity.

UNDO

Don't be sedentary—stand up and move throughout the day.

Examine your negative emotions.

Heal injured relationships that are meaningful to you.

Be mindful of lapses and imbalances in your diet.

Address negative stereotypes about aging and ageism.

Consider how to heal the fear of death.

The good news about preventing or even reversing the aging process is that it has become realistically possible. The time of wishful thinking has passed. Increasingly, the medical community knows what we're dealing with as the body ages, which wasn't true before—in fact, aging was quite a mystery. There is no single process known as aging. Instead, aging is as multidimensional as life itself. To hear that aging is almost impossible to define will surprise most people. They identify aging with its symptoms—loss of muscle mass, wrinkles, fading vision, and so on. But the symptoms of a cold aren't the same as knowing the actual cause, and the symptoms of aging are as far removed from the cause as a runny nose is from the cold virus.

Current research has zeroed in on genetic changes as the key, and genetic activity, as we've seen, can be strongly influenced by lifestyle.

Since people are living longer, it's realistic to say that past age fifty, you will enter a second lifetime, and unlike children, who spend the first two decades of life occupied with developing into fully capable human beings, a fifty-year-old can bring a wealth of knowledge, skill, and experience to the second lifetime that now presents itself. In a word, how you age today—or don't—will turn old age into a rising arc or a steady decline. Despite the influence of genes and biology, the choice is largely yours.

As things stand, the universal experience of growing older can't be reduced to a single cause or a single result. What society believes about aging and the elderly can be just as important as what's going on biologically. The adage "You're only as old as you think you are" points to a third factor, the psychological. Taken altogether, the picture of aging has been confusing, leading to a collection of basic facts that apply differently to each person, as follows:

- Formerly, aging was thought to begin biologically starting around age thirty and proceed at around 1 percent of physical deterioration a year for the rest of a person's life. Now we realize that this view was tied to symptoms of aging. At the cellular and epigenetic level, signs of impaired function can, and do, begin much earlier.

- The whole bodymind system is affected by the aging process, but not at a predictable rate.

- Because the aging process is so variable, some people are biologically younger than their chronological age, some older.

- Aging leads eventually to death from a specific breakdown in one system (usually the respiratory system). At the time of death, the vast majority of cells are still functioning normally, or at least well enough to keep the person alive.

- For every typical sign of aging, there are at least a few people who improve as they grow older, including in such areas as memory, muscle strength, and mental acuteness. This raises the possibility that aging may not be necessary. If that's true, why do we age?

Faced with such a confusing picture, medical science couldn't make aging fit the disease model—aging isn't the same as getting sick, even though the elderly are more prone to illness than young adults. The holy grail in physics, sought after for many decades, is known as the theory of everything, a unified explanation of all the fundamental forces in the universe. In medicine there is no comparable theory of everything in regard to aging. When you catch a cold, you will exhibit symptoms over the course of a week typical of almost everyone else who catches a cold, but aging takes decades to unfold, and no two seventy-year-olds are exactly or even approximately alike. You are a unique person, and aging highlights your unique qualities.

Anti-aging took a leap forward in the last two decades when it became clear that the aging process is centered on DNA. Now we know, thanks to the field of epigenetics, that a lifetime of experiences constantly affects gene activity, leaving marks or imprints that last a long time. No one can say with certainty if a specific marker lasts for years, decades, or a lifetime, but the crucial fact is undeniable: Your lifestyle has genetic consequences. Even identical twins, born with the same genome, will exhibit gene activity in their seventies that is as different as two siblings who aren't twins; sometimes the differences will be the same as between two total strangers.

The most recent breakthrough in anti-aging is the finding that the aging process begins young. In a 2015 Duke University study headed by Daniel W. Belsky, the focus was on biological age (how old your body is) as opposed to chronological age (how old you are by the calendar).

Traditionally, the aging process has been studied primarily in old people, who already show symptoms of lifestyle disorders. Instead, the Duke team looked at 954 young people, tracking the biomarkers for aging at three different time points between ages twenty and forty: "Already, before midlife, individuals who were aging more rapidly were less physically able, showed cognitive decline and brain aging, self-reported worse health, and looked older." This finding helps boost the idea of anti-aging by pushing the whole issue back decades before the signs of disease and infirmity are advanced. As we've been demonstrating throughout this book, a long trail leads to the start of many disorders, and now aging, which affects every system in the body, joins the list.

Finding the most reliable biomarkers for aging is still controversial—the possibilities range from deep neural networks to T-cells and epigenetic markers. Only when this issue is settled can anti-aging be measured with confident accuracy. Right now there is no gold standard for measuring aging. This isn't a surprise, given that the process is incredibly complex and affects each person differently. Yet by any standard, the burden of anti-aging rests with each individual, not with the promise of a silver bullet in the future.

As you make your choices today, keep one thing in mind. Just as you are unique in the way you age, anti-aging will also be unique. The more you understand the aging process, the better you can individualize your anti-aging program. Here are the most important variables that affect aging in general, according to the best recent research.

Successful Aging: Top Ten Variables

1. Satisfying relationships with family, friends, and community

2. Emotional resilience, the ability to bounce back from setbacks and losses

3. Stress management

4. Anti-inflammation, including diet and "inflamed" emotions like anger and hostility

5. Good sleep every night

6. Meditation, yoga, mindful breathing

7. Moderate physical activity throughout the day. Standing up and moving to break up long periods of sitting.

8. Positive attitudes toward aging and the passing of time

9. Absence of toxins, including tobacco and alcohol

10. Youthful mindset—curious, open, always learning new things

Listed roughly in the order of priority, these variables provide insight into how people age well. We should note, however, a cutting-edge theory of aging that holds inflammation responsible for every aspect of the aging process. Although unproven, this theory may be the wave of the future, considering that so many lifestyle disorders basically occur in old age and at the same time are linked to low-grade chronic inflammation.

As with every aspect of a healing lifestyle, waiting until you show visible symptoms of aging comes too late. Aging is the ultimate example of a creeping, incremental change surrounded by a litany of influences. Anti-aging is also incremental but clear in its strategy: maximize the positive input your bodymind receives every day and minimize the negative input. *Input* is an all-embracing term, but the best research points to the areas everyone should focus on, which is what our "Do" and "Undo" recommendations are about.

The "Do" Side

We present only one age-specific choice on the "Do" list—taking a multivitamin and mineral supplement if you are over sixty-five (see page 184 for why this choice is important). The other choices pertain to your welfare and happiness right now, based on the notion that a happy life is built upon being happy every day. The longest-running study on aging is the Harvard Study of Adult Development, now eighty years old, which arrived at findings summarized in a headline at the website Harvard Gazette: "Good genes are nice, but joy is better." The study began in 1938 with the intention of following 268 Harvard sophomores throughout their lives. (The subject population was later expanded and diversified. Only 19 of the original subjects were still alive in 2017, but their

1,300 offspring are being studied, along with their wives and additional inner-city volunteers.)

Robert Waldinger, a psychiatrist and professor at Harvard Medical School who now heads the study, reports: "The surprising finding is that our relationships and how happy we are in our relationships has a powerful influence on our health. Taking care of your body is important, but tending to your relationships is a form of self-care too. That, I think, is the revelation." This finding fits into previous points we've been making about heart disease, for example, where answers on social support or a spouse who expresses love turn out to be good predictors of who will exhibit symptoms of heart problems and who won't. At the opposite extreme, to quote Dr. Waldinger: "Loneliness kills. It's as powerful as smoking or alcoholism."

These findings aren't tentative or confined only to an upper social group. As the article on the Harvard study goes on to say: "Close relationships, more than money or fame, are what keep people happy throughout their lives, the study revealed. Those ties protect people from life's discontents, help to delay mental and physical decline, and are better predictors of long and happy lives than social class, IQ, or even genes. That finding proved true across the board among both the Harvard men and the inner-city participants."

Our "Do" choices focus on this key finding. The more social support you have and the happiness you find in relationships will continue to affect you throughout your lifetime. In the "old" old age, as outdated attitudes are labeled, the golden years were a euphemism for being stuck in a rocking chair and being of no use to society once you passed sixty-five. In addition, people looked forward to being happy after they retired, setting this as their main goal rather than being happy here and now. The ethos dictated that you worked as hard as possible during your prime years and postponed happiness until you retired; it was one of the perks, so to speak, of not having to work. In the "new" old age, a set of attitudes still evolving, chiefly among baby boomers, there is no intention of retiring as long as one's work is useful and fulfilling. In aid of this, people intend to preserve their healthy status for as long as possible, preferably until the final illness.

Where the "new" old age needs to grow is in the area of social support and relationships, because happiness for too many people is still an individual project. The ethos of American individualism stands at the opposite pole from a communal society like Japan or countries with policies of social welfare, like almost every European nation. In our list of "Do" choices we present meditation as something where going it alone seems mandatory, but even there, people who join meditation groups are more likely to keep up their practice.

The most valid measure of quality of life is how happy you are, how fulfilled and contented your lifestyle has made you. People who spend their careers becoming secure financially often lack any but rudimentary skills in sustaining a happy relationship. We can't address this problem in full—that would take ten times more space than we can devote to it—but Deepak's book *The Ultimate Happiness Prescription* makes the following points:

- Happiness is hard to predict. People think that they will be happier when they have more money, a baby, a job promotion, and other external factors, but there is no correlation between these expectations and actually being happier. While having enough money and security is an important component of being happy, beyond that point making more money doesn't increase happiness and often has the opposite effect of adding more stress to a person's life.

- Because it is so unpredictable, happiness should be addressed today rather than postponed for the future.

- Each of us has an emotional set point, like the body's metabolic set point, that chiefly determines our mood from day to day. After an unhappy event, whether it's a bad breakup or a financial loss, we return to our set point, generally within six months.

- Even taking this set point into consideration, the current psychological understanding is that at least 40 to 50 percent of happiness depends on lifestyle choices.

- In the world's wisdom traditions, the fickle nature of human happiness cannot be solved by seeking happiness externally. Only by finding a level of the mind that is established in inner peace and contentment can the problem of unhappiness be solved—this is in keeping with our chapter on the end of suffering.

The "Undo" Side

The choices offered on the "Undo" list revolve around a central theme: getting unstuck. You have to be as resilient in your approach to life as your cells are. If you are rigid in habits, behavior, and attitudes, you will incrementally decrease the ability of your cells to thrive and remain strong in the face of challenges. Remember, the bodymind is a single process operating with hundreds of subprocesses twenty-four hours a day. No experience goes unheeded. Clenching your mind is the same as clenching your fist—at a certain point you will develop a cramp.

Begin today to look at the negative attitudes that tend to increase with age if you aren't mindful of them. These include the following:

"Growing old is horrible. It's all downhill from here."

"The prospect of death is terrifying."

"The best years are behind me."

"The past was a lot better than today."

"You can only look out for number one."

"People always let you down."

"Time is running out."

These attitudes and beliefs are not testable against reality one way or another—they are held not for factual reasons, but for emotional ones. The whole point is how you choose to feel about your life and the future. If you hate and fear the thought of growing old, aging will become more and more negative as time passes. Every new sign of growing older, from graying hairs to pains in your joints, will be another reason to hate and fear where life is taking you. Limiting belief systems are the major obstacles to healthy aging. It is important in mind, body, and spirit to always be looking forward to something, today and tomorrow, while not obsessing about the past.

Since every belief is a personal creation, it can also be uncreated. We devote Thursday to core beliefs and how to change them, but for now, the undoing process involves a few mindful steps:

- Associate with inspiring, happy older people, beginning today.

- At the same time cultivate connections with young people.

- Don't participate in conversations in which people complain about aging.

- Every negative belief about aging can be countered by consciously replacing it with a positive belief, as follows:

  - *"Growing old is horrible. It's all downhill from here."*

  - Replace with "My life is a rising arc. The best is yet to come."

  - *"The prospect of death is terrifying."*

  - Replace with "Fear never solves anything, including this."

  - *"The best years are behind me."*

  - Replace with "I can create a better future if I choose to."

  - *"The past was a lot better than today."*

  - Replace with "Dwelling on the past cancels out the possibilities of today and tomorrow."

  - *"You can only look out for number one."*

  - Replace with "All my life I have looked out for others, and they have looked out for me."

  - *"People always let you down."*

  - Replace with "People fundamentally are doing their best."

  - *"Time is running out."*

  - Replace with "There's always enough time."

Since beliefs are held for emotional reasons, we aren't saying that positive beliefs are always factually true, only that your emotional state is where the most powerful motivations lie. That's a very significant part of aging successfully. Having a positive attitude toward it makes an enormous difference given that decades of living are involved. Positive thinking, however, tends to be superficial and therefore not as important as self-acceptance. When you have that, even the worst indignities of old age—which of course we want you to avoid—don't turn into a downward spiral. A strong sense of self weathers any storm.

The Telomere Connection

Now that it's accepted that people age differently, it's critical to know why. Aging is such a holistic process that you might suppose that no simple answer about why we age differently will ever emerge. But that may not be true at the cellular level. Cells have their own lifespan, ranging from the early stages, marked by rapid division and fresh renewal every time a cell divides—this is the period molecular biologist Elizabeth Blackburn calls luxuriant growth—ending with a stage in which no more divisions occur and the cell is tired and unreliable in performing its basic functions—this is the period known as senescence.

A senescent cell breaks down on several fronts. It sends out faulty chemical messages and fails to interpret incoming messages correctly. Its ability to heal itself slows down and eventually comes to a halt. Pro-inflammatory substances can start to leak out through the cell membrane into surrounding tissue and the bloodstream. It seems more and more possible that when our cells age, so do we.

The most striking support for this theory comes from research into our genes and specifically a section of DNA known as a telomere, which caps the end of each chromosome like a period ending a sentence. Telomeres are "noncoding" DNA, meaning that they have no specified function in building cells, but they are far from passive. Their function seems to be to preserve cells. Every time a cell divides, which happens constantly somewhere in the body, its telomeres are shortened. Longer telomeres are typical of young cells in the stage of luxuriant growth; shortened or frayed telomeres are typical of weary senescent cells.

The leading researcher on the subject is someone we've already briefly mentioned—molecular biologist Elizabeth Blackburn, who shared the 2008 Nobel Prize in Physiology or Medicine with Carol Greider of Johns Hopkins and Jack Szostak of Massachusetts General Hospital for their discovery of telomerase, the enzyme that replenishes telomeres. Now the head of the Salk Institute in La Jolla, California, Blackburn covers every aspect of cell aging and renewal in her 2017 book, *The Telomere Effect*, coauthored with her close research colleague of fifteen years, health psychologist Elissa Epel. They convincingly describe telomeres and levels of telomerase in the cell as our best marker yet for the mysterious and multifold process of aging. This also implies that by increasing one's telomerase levels and thereby causing telomeres to grow longer, a healthy lifespan can be founded on cells that keep renewing themselves for decades.

In their book, Blackburn and Epel cite a startling actuarial prediction. There are currently around 300,000 centenarians existing around the world, a number that is rapidly increasing. According to one estimate, reaching age one hundred is about to become so commonplace that one-third of children born in the United Kingdom will live to be centenarians—the issue of protecting your cells is suddenly more urgent than ever. We highly recommend reading Blackburn and Epel's book— its wealth of information needs to be absorbed in detail. But the bottom line is to understand what puts your telomeres at high risk and low risk.

The book's survey of all the pertinent research dovetails with everything we've been discussing under a healing lifestyle, as follows:

Your telomeres are at low risk if you
- *Have no exposure to severe stress.*
- *Have never been diagnosed with a mood disorder.*
- *Enjoy good social support, including a close confidant who gives good advice, friends who listen to you and with whom you can unburden yourself, and relationships in which love and affection are shown.*
- *Exercise moderately or vigorously at least three times a week, preferably more.*
- *Get good-quality sleep for at least seven hours a night.*
- *Consume omega-3-rich food three times a week while avoiding processed meats, sugary sodas, and processed food in general. A whole-food diet is best.*
- *Are not exposed to cigarette smoke, pesticides, and insecticides.*

The opposite is also true.

Your telomeres are at high risk if you
- *Are being exposed to severe stress in your life.*
- *Have a history of being medically treated for anxiety or depression.*
- *Lack social support from friends and family.*
- *Lead a completely sedentary lifestyle with no regular exercise, even light activity like walking.*

- ~ *Suffer from chronic insomnia or cut your sleep shorter than seven hours a night.*
- ~ *Consume a diet high in fat, processed foods, and sugary sodas, with no attention to sufficient fiber and omega-3 fatty acids.*
- ~ *You are exposed to cigarette smoke, pesticides, insecticides, and other chemical toxins.*

These points summarize the research-supported risk factors presented by Blackburn, and as with any risk-based program, some people are more affected than others. Severe stress is one of the most thoroughly damaging factors—in one study, caregivers who tended Alzheimer's patients had shortened telomeres that predicted a shortened lifespan of between five and eight years. Blackburn also lists a number of commercial labs where people can pay to have their telomerase level analyzed.

It's also significant that the lifestyle choices known to decrease the risk of heart disease, particularly the intensive one devised by Dean Ornish (see page 53), have a beneficial effect on telomere length. Extending the program to cancer, Ornish had another impressive finding. A group of men with low-risk prostate cancer were selected for study. (Low risk means that their cancer was at an early stage and slow growing. Prostate cancer can take decades to advance, and the current recommendation advises balancing the risk and reward of doing any active treatments, a change from the era when any cancer was immediately treated and usually aggressively.)

The men were put on a variant of the heart-disease protocol: they ate a low-fat, high-fiber diet; walked for thirty minutes a day; and attended regular support group meetings. Stress management was included, and there was training in meditation, mild yoga stretching, and breathing. At the end of three months the group that was on the program had higher telomerase levels than the control group, which meant that their cells were aging better. Stress seemed to play a key role, because the greatest increase in telomerase occurred among the men who reported having fewer distressing thoughts about prostate cancer. Ornish followed some of the men for five years, and those who stuck with the program showed telomeres that had increased by 10 percent, reversing the usual expectation as cells age.

If stress levels determine how well or badly our cells age, this should show up in meditation studies, and it has. Blackburn cites two studies conducted at meditation retreats that were three weeks and three months long. At the end of the three-month retreat, meditators had higher telomerase levels compared with the control group. In the three-week retreat meditators showed longer telomeres in their white blood cells than the control group, which showed no change.

How long would it take for these effects to appear, and how intensive does your dedication to meditation have to be? There is no definitive answer, but probably the best clues come from a collaborative study that we carried out with Blackburn and other leading researchers, conducted at the Chopra Center in Carlsbad, California. Women in good health were divided into two groups. One group enjoyed a spa vacation without other interventions. The other group went through a program led by Deepak that included meditation and a variety of Ayurvedic treatments. At the end of the week everyone reported feeling better, which attests to the likelihood that most people are in sympathetic overdrive, because simply going on a week's vacation improved their sense of well-being.

In the same vein, there were improvements in gene expression (activity) in both groups, including the chemical pathways that trigger inflammation and the stress response. There also appeared to be a meditation effect on telomeres and telomere-protective genes. These occurred in the meditation group among experienced meditators. The fact that it took only a week to produce results that started to be significant points to the conclusion that you are doing your cells some good almost as soon as you take up meditation and that the practice needs to be regular and long lasting.

We are encouraged by how strongly telomere research validates the healing lifestyle this book advocates. It also underlines the conviction that cells benefit directly, at the genetic level, from conscious lifestyle choices. Blackburn ends her book with a visionary Telomere Manifesto that would prioritize protecting our cells as a part of parenting, social relations, the fight against income inequality, and global outreach for the planet. Like all visions, this one depends upon individual decisions, and one comes away from *The Telomere Effect* even more persuaded that anti-aging begins by keeping our cells in a state of renewal. If there are no new startling things to do, becoming more optimistic about your own aging is valuable in itself.

Stand, Walk, Rest, Sleep

Today's recommendations—choose only one.

DO

Stand up and move around once an hour if you are working at the computer or at a desk job.

Walk 5 minutes for every hour you work.

Take the stairs instead of the elevator.

Park your car far away in the lot when you shop or go to work.

Be regular in your sleep routine.

Make your bedroom an optimal sleeping environment (see page 228).

Walk for 20 to 30 minutes in the evening.

Take 10 minutes of quiet alone time, preferably in meditation, twice today.

Spend more time with a physically active friend or family member.

UNDO

Replace 10 minutes of sofa time in front of the TV with a walk instead.

Break the habit of waiting until the weekend to catch up on lost sleep.

If you drink alcohol, do it early in the evening—go to bed without alcohol in your bloodstream.

Replace the midmorning coffee-and-doughnut break with a walk.

Walk to one place close by that you usually drive to.

Examine your excuses for not being more active.

Not getting enough sleep bothers many people, but it can't be addressed alone. Today's theme is expanded to include the complete cycle of rest and activity that benefits the bodymind. As a society, we've created a situation with sleep that works against the biorhythms governing the whole system. If you sit all day and get no significant exercise, you can wind up "too tired to go to sleep," because the rhythm of sleep and activity has been disturbed. Research has shown how interlocked our needs for rest and activity actually are. To keep your biorhythms synchronized, four elements must be present:

Standing: Simple as it sounds, the human physiology depends on gravity. Seminal research in the 1930s showed that college athletes, when confined to bed for two weeks, lost months' worth of muscle tone in their training. Standing up for only a few minutes a day keeps muscle tone intact. It also appears to aid in recovery from surgery, which is why patients are no longer advised to get constant bed rest in the hospital but encouraged instead to stand up and walk if they are able.

Walking: Although exercise delivers more benefits the harder and more frequently you exercise, the baseline for activity is walking. Research has shown that the widest gap in levels of physical activity, medically speaking, occurs between those who take zero exercise and those who get up off the sofa and do something, no matter how meager. Walking is now a regular practice in recovery from serious illnesses and surgery.

Rest: After heavy physical exertion, rest is necessary to replenish your muscles and restore internal balance—most people have no difficulty with this because they feel exhausted after heavy work or exercise. But the need for mental rest has only recently been taken seriously. If you equate mental rest with lethargy and dullness, that image is misleading. People who practice meditation, which among other things rests the mind, emerge with sharper alertness. Meditation doesn't dull the mind or put the brain to sleep—there is actually increased brain activity (in alpha waves, for example, which are associated with creativity), resulting in a state previously unknown to neuroscience: restful alertness.

Sleep: Researchers still don't know why we need to sleep at all, except that undeniably we do. The most recent theory is that sleep allows the brain to rid itself of built-up toxins during the day. These include, during the deepest stage of sleep, the removal of senile plaques that can cause Alzheimer's disease. It is also during deep sleep that we consolidate what we have learned all day as short-term memories into long-term memories. Without these activities, our brain (as well as the rest of our body) can undergo damage done by lack of sleep and poor sleep.

Let's do a deeper dive. The first thing everyone notices when they spend a sleepless night is feeling tired and groggy in the morning, sometimes throughout the day. This becomes a chronic complaint from insomniacs, yet even when someone says, "I didn't sleep a wink last night," dream studies reveal that intermittent episodes of sleep do in fact occur, even though they may be fitful and shallow. If someone is forced to stay awake the entire night, for example, in a sleep clinic laboratory, more serious deficits begin to show up, such as lack of motor coordination and attention—these are serious causes of motor vehicle accidents. Chemical imbalances start to show up, particularly in the flow of hormones, which are precisely balanced according to our circadian (daily) clock. Not getting enough sleep can disturb your appetite because the balance of leptin and ghrelin, the two hormones that govern hunger and satiation, has been thrown off.

Except in a sleep lab, few people go beyond losing one night's sleep; the brain's demand for sleep is too hard to overcome. But extended lack of sleep leads to headaches, muscle weakness, tremors, hallucinations, and other serious symptoms. Because these drastic effects aren't experienced doesn't mean, however, that you aren't suffering from sleep deprivation. As with low-grade chronic stress and inflammation, the habit of losing sleep creates problems incrementally over the long run. Insomniacs run a higher risk of anxiety and depression, for example. Knowing this, psychiatrists will caution their patients who suffer from chronic depression to be on the lookout for poor sleep. This has been shown to be one of the earliest signs that a bout of depression is about to begin; it also marks an early enough stage that the onset of depression can sometimes be averted simply by correcting the irregular sleep pattern. Use of drugs like cocaine will often lead to poor sleep, which then leads to depression and anxiety, stoking a desire for more drugs in a vicious circle.

In a 2003 review article in the journal *Behavioral Sleep Medicine*, the authors reported wide-ranging psychological effects: "Insomnia was consistently predictive of depression, anxiety disorders, other psychological disorders, alcohol abuse or dependence, drug abuse or dependence, and suicide, indicating insomnia is a risk factor for these difficulties." This finding has its milder implications, as anyone who has lain awake worrying knows all too well. The review article goes on to note that insomnia is associated with decreased immune response, while the data were inconclusive over whether insomnia is a risk for cardiovascular disorders. Sleep studies tend to be small, and the definition of insomnia is fairly ambiguous, but it sounds disturbing that the regular use of sleeping pills is a risk for mortality (i.e., a shortened lifespan). A 2012 study at Scripps Health in San Diego linked popular prescription sleeping pills to a five-fold increased risk of early death. This study indicated increased risk of death for both casual and heavy users.

One of the usual suspects, inflammation, also seems to enter the picture. In a 2010 study, subjects were kept awake for twenty-four hours or more and showed increased markers for inflammation (cytokines). The results weren't significant enough to be called clinical—needing medical treatment—but it's notable that increased inflammatory markers also showed up in subjects who were awakened after only two to four hours of sleep. No reliable cause could be found for the increased inflammatory markers, but speculation pointed to "autonomic activation and metabolic changes," which we simplified as sympathetic overdrive (see chapter 5). In other words, the sympathetic nervous system was put under stress.

The stress and strain of modern life keeps the sympathetic nervous system chronically stimulated. When you toss and turn, unable to get to sleep, you may blame the thoughts running through your head, or physical tenseness and tightness, or a nameless refusal of the sleep response to click in. But these diverse symptoms are generally traceable to being in autonomic overdrive. The stress response has been subtly activated, and one effect is to be alert—that's part of responding to external threats. In acute stress the pupils dilate, heart rate rapidly increases, and the surge of adrenaline demands action, either fleeing or fighting. Low-grade stimulation of the stress response isn't so dramatic, but at any level, the stress response negates falling asleep. Stress and sleeplessness form a vicious circle, and if you are also stressed out about your insomnia, the effect is only aggravated. Our recommendations about stress reduction will go a long way to breaking the stress-insomnia connection.

The "Do" Side

Babies and young children fall asleep effortlessly. Being tired from a physically active day also makes sleep automatic. But most of us expend fewer and fewer calories per day in physical activity. Studies have shown that sitting at the computer uses about 80 calories an hour. You can use 8 to 10 more calories by walking for five minutes four times an hour, and over the long run, this is enough to control the slow, gradual weight gain that plagues people as they age. (Over an eight-hour workday, adding just 10 calories an hour, if regularly kept up, would offset 20,000 calories a year, or around 6 pounds of weight gain.) Standing desks are gaining in popularity, and they have some health advocates. However, standing adds only 2 calories an hour over the energy expended sitting down.

The trend toward using fewer calories is likely to increase in the future, which eliminates the easiest way to assure sound sleep. Therefore our recommendations focus on some basic lifestyle changes you can adopt for life. You may wonder why we haven't included the standard governmental recommendation of moderate-to-heavy exercise lasting at least thirty minutes three to five times a week. The answer is compliance. Studies have shown that Americans are exercising a little more than in the past, but the age group that takes exercise is young, nineteen to twenty-nine years old, with steady declines every decade thereafter.

The least active group are seniors, and that needs to reverse. Longevity and good health in old age increase with activity and decrease among people who give up and sit all day. Everyone over age seventy in good health would benefit from mild cardiovascular workouts and mild weight training, even into their nineties. For the sake of compliance, the secret is to habituate yourself early. Being regular in simple things like standing and walking makes you much more likely to keep up the habit as you age. For mental alertness, meditation is recommended because you get to experience the state of restful alertness. As we pointed out above, this is a state of awareness that is neither asleep nor dull. In restful alertness the mind is fully awake but not stimulated. Everyone at every age group will benefit from making this state a familiar habit.

Your sleep environment: Here's a checklist for turning your bedroom into an ideal place for sleeping.

12 Steps to Getting the Best Sleep

1. Make the room as dark as possible, using blackout shades or wearing a sleeping mask.

2. Make the room as quiet as possible.

3. If you share a bed with someone who snores, wear earplugs.

4. Avoid using your bed for work.

5. Don't text in bed.

6. Keep your room on the cooler side.

7. Stop watching TV at least an hour before bedtime.

8. Keep the TV in another room.

9. Make the room as soothing to the senses as possible with colors and scents—this should be the room in your house you associate with relaxation.

10. Buy a comfortable mattress with sufficient back support—the firmer the better for most people.

11. Use a hypoallergenic pillow.

12. Wash bedclothes frequently enough to eliminate dust.

The first imperative is a completely dark, preferably pitch-black bedroom. There's a physiological reason for this. The pineal gland, which is buried deep in the brain, is crucial to regular sound sleep because it is sensitive to light. As you sleep, brain activity fluctuates, and at the end of seven to eight hours, you rise toward wakefulness in waves. You aren't aware of being awake until the last wave lifts you out of sleep, yet if your bedroom is lit with morning sunlight, you will tend to wake up too early, on a shallow wave. This interruption can be fairly easy to overcome—you bury your head in the pillow and go back to sleep. But because you didn't get seven to eight hours of *continuous* sleep, you will usually feel groggy once you are fully awake. (If you're a frequent traveler, you might notice how well you sleep in hotel rooms. That's because they have blackout curtains that make the room considerably darker than the average bedroom.)

Blocking out extraneous noise is important for two reasons: it can keep you from falling asleep and also wake you up too early (on one of those shallow waves). Another recommendation aside from improving your sleep environment is to take your daily low-dose aspirin—recommended for all adults as a preventive for heart attacks and even some forms of cancer—at night. As we mentioned earlier, minor aches and pains that go unnoticed during the day can become irritants when you get into bed. Taking an aspirin helps to eliminate this often-overlooked contributor to insomnia.

The "Undo" Side

If noncompliance is the obstacle on our "Do" list, inertia is the enemy on the "Undo" side. Habits are self-reinforcing. If you skip a day of exercise, for example, it's easy to let the next day slide. Yet for every skipped day, you lose the benefit of a day of exercise, so inertia leads steadily downhill. (As a side note, this pattern pertains to who has a good sex life as they age. The people who are most likely to have satisfying sex are the ones who didn't stop. Having sex reinforces itself; not having sex reinforces itself.)

We believe that avoiding inertia rarely works if someone adopts a strenuous habit like running every day—for every hundred people who adopt this routine, few will maintain it for life. The day comes when you quit running, and then there's a steep decline before you reach the place occupied by people who don't run and never have. Yet the example of brushing your teeth every morning reflects how easy it is to adopt very simple, undemanding habits.

Over a lifetime, if you follow the pattern of stand, walk, rest, and sleep every day, you will do a great deal to stay in optimal health. Our "Undo" recommendations gently nudge you to avoid the slow trend toward inertia.

One reason that people give for not exercising when they took it up to lose weight is that exercise didn't work. Not only did they not lose weight, but because exercise is physical work, so far as your metabolism is concerned, they gained weight by being hungry. (In regular strenuous exercise, like training for a marathon, there can be weight gain because the training regimen replaces fat with muscle, and muscle weighs more than fat. Of course, a marathon runner's heavier body is likely to look more attractive than the body of a couch potato of the same weight.)

For a long time the complaints about not losing weight were ignored, until genetics revealed that some people are biologically disposed to increase their metabolism during exercise, which burns calories, while other people are not so disposed. As a system approach would predict, genes don't tell the whole story. What you eat and how you eat also affect metabolism, as do your stress level and the hormones controlling hunger and satiation. The cloud of causes is prevalent once again.

Leaving weight aside, even though it's one of the biggest reasons that motivate people to exercise, along with looking more attractive, being physically active also has varying effects. At one extreme are those who get the famous runner's high; at the other extreme are people who just get very tired when they run. Strenuous exercise is associated for some with being good at sports, a positive reinforcement. There's no positive reinforcement, however, if you hated gym class and never made the team in a sport.

The upshot is that how you feel about being active should dictate your choices. There is no one-size-fits-all exercise program, despite the benefits promised by the governmental guidelines about regular exercise. Our concern is to cross the gap that separates a totally sedentary life from an active one. Our solution is the stand, walk, rest, sleep formula. If you can go beyond this stage, more power to you. But be aware of one important thing. Stand, walk, rest, sleep isn't a bare minimum that marks you as a slacker. Instead, it's a healthy norm that you can maintain for life even when the winning high school quarterback has developed a pot belly along with his beer-drinking buddies.

Core Beliefs

Today's recommendations—choose only one.

DO

Write down five core beliefs and evaluate why you believe them.

Put a core belief into action.

Read a poem, scripture, or spiritual passage for inspiration.

Have a family discussion about which core beliefs everyone holds.

Take your favorite role model and list what core beliefs they held.

UNDO

Examine your negative beliefs as they relate to fear and mistrust.

Open a line of communication with someone who holds radically different values.

If you are stuck on a negative belief, be a devil's advocate and argue against it.

End your participation in us-versus-them thinking.

Today is about your deepest beliefs, the ones you identify with over a long stretch of time. They can have healing value or the opposite, because beliefs get turned into thoughts, words, and actions that the body-mind reacts to. Everyone holds on to personal beliefs, and in one way or another we are emotionally attached to them. But not all beliefs are created equal. Some are just opinions, picked up and dropped without much effort. Other beliefs are secondhand attitudes we absorbed along the way, usually beginning in childhood with our parents' belief system (as in the choice of religion). Studies find that 70 percent of first-time voters vote for the same party as their parents, and from then on, they tend to stick with their choice.

These sorts of beliefs are at best incidental to a healing lifestyle, but at a deeper level your health and well-being are strongly influenced by what we're calling core beliefs. Core beliefs lock in your perspective on some crucial questions:

Is life fair?

Is there a higher power in the universe?

Does good triumph over evil?

Should I expect the best or prepare for the worst?

Should my attitude be relaxed or vigilant?

Am I safe?

Am I loved, cared for, and supported by others, or can I count only on myself?

Am I good enough and smart enough?

How your life turns out depends on your answers to these questions. In modern times the responsibility for answering them rests on the individual. Whether or not you are consciously on a spiritual path, you have been seeking and finding answers to higher questions all your life. In an age of faith, by contrast, fixed and authoritative answers were provided by religion. Here we are concerned not with the philosophical side of core beliefs, but with how they affect the bodymind. If you feel unsafe in the world, for example, your life will be psychologically very different from that of someone who feels safe, and depending on how much threat you perceive, you may experience much more stress.

We've already discussed the possibility that an infant's worldview may be genetically determined at the epigenetic level (page 163), which would be disturbing if the infant's programmed view of life was painful and discouraging. However, it's almost certain that a familiar term, a *cloud of causes*, is at work. We form our core beliefs in a fog of influences, and for countless people the fog never clears. Take the first question on our list, "Is life fair?" and compare how two people might hypothetically arrive at opposite answers.

Person A has heard over and over that life is unfair, and he accepts the truth of this. Looking around, he sees good people being hurt while bad people rise and are never punished. Thinking back on his own experiences, lots of times there were unfair outcomes—the girl he loved and lost, the promotion he was passed over for, the deal that collapsed because someone backed out at the last minute. The news is full of unsolved crimes and biased jury decisions that allow the guilty to go free. Who could possibly claim, just looking at the gross inequalities in the world, that life is remotely fair?

Person B hasn't led a charmed life, but she's had no major setbacks, and from that perspective life has been more than fair—it's been abundant and generous. Loved as a child, she married someone she loved and made the right choices. Their children are healthy and happy. B knows that there is ugliness and unfairness in the world, but her Catholic faith tells her that God alone is judge and God acts in mysterious ways. It's up to us to accept that God created a benevolent universe and that mankind can be redeemed from sin. This overarching view outweighs the weakness and badness that human beings fall into.

A and B hold opposite beliefs for all kinds of reasons, and there is no mathematical equation for how to weigh each influence, because the cloud of causes changes over time. We can't concern ourselves with whether A or B is actually right, because core beliefs never match reality with a capital *R*; beliefs, as we mentioned earlier, are all about your personal reality. Yet there are core beliefs that support health and well-being and those that don't. Here are the relevant factors.

A Belief Is Healing If . . .

It's flexible, tolerant, and open to change.

It promotes happiness.

It is loving and kind.

You gain in self-esteem from it.

It doesn't create stress for you or other people.

You don't use your belief to fuel anger, fear, and mental agitation.

It helps you bond with family, friends, and community.

It encourages an optimistic outlook.

As you can see, we are using a broad definition of healing, which is justified by a whole-system approach. For many people there's a vague sense that being positive is better than being negative, but we aren't promoting positive thinking—we're promoting a healing attitude toward the self. A core belief that leads to inflammation or a stress response is like bad input into the body's information superhighway. The difference between inflammation caused by cutting your hand and inflammation caused by getting enraged at a political story on the evening news is minimal from the viewpoint of a cell that must contend with pro-inflammatory markers in the bloodstream.

In an earlier chapter we recounted the amazing recovery by Norman Cousins from a potentially fatal illness. After he became a crusader for the mind-body connection, Cousins had a story he liked to tell that shows the power of belief in the bodymind. He spotted the story in the *Los Angeles Times* in 1983, about an outbreak at a local high school football game. Four people became ill with symptoms of food poisoning during the game and were attended by a doctor on the scene. It turned out that all four had drunk Coca-Cola from a fountain dispenser, the kind that mixed the Coca-Cola syrup with soda water.

The doctor didn't know if the contamination came from the water or the syrup, and since copper pipes were involved, there was also the possibility of copper sulfate poisoning. An announcement was made warning the crowd not to drink the Coke, and within minutes 191 people fell sick enough to be hospitalized. Several hundred more started retching or fainted; many rushed home to contact their family physician. As Cousins commented, "[I]f we just hover over this a bit, what we come up with is the fact that sounds in the air can be translated into specific physical illness. Those symptoms were not feigned. [They] were real, as anyone who watched those people retching could testify."

The same invisible trigger can come from inside us, creating a pathway from belief to inflammation and stress or actual disease symptoms.

No one willingly desires food poisoning, so why do we tolerate self-created damage to the bodymind? One factor is what psychologists call *secondary benefit*. This is a psychological mechanism for offsetting pain, like the lollipop offered to a child after getting a vaccination. Another example is severance pay, which is meant to soften the blow of being fired. At first sight, secondary benefit looks like a useful mechanism for coping with pain and misfortune, but when misused, it turns into self-defeating tactics like denial.

When a negative state lasts long enough, we become desperate to find a way to cope. Chronic anxiety is a perfect example. It has recently come to light that anxiety is a serious problem among the young, down to ages that no one ever previously associated with anxiety. But in reality, anxiety can become chronic as early as age four, and it is now known that every serious mental disorder in adulthood is associated with anxiety in childhood.

Disturbing as this is, the reason childhood anxiety remained hidden for so long, even from professional therapists, is that children are good at finding ways to disguise it, even from themselves. They shove the feelings down; compensate with play and distractions like watching TV to divert attention from their fear with other negative behaviors like bed-wetting; or simply learn that Mommy and Daddy don't want to hear about these feelings. Anxiety is intolerable as a day-to-day experience, so the mind must find an escape, however ineffectual.

Much of this is unconscious behavior we turn into mindless habits. Consider a self-defeating habit like hating the opposite political party or turning your annoying neighbor into an enemy. Why would you cling to such a negative attitude, even knowing that it isn't healthy? The reason boils down to mindlessly reinforcing your reaction over and over again instead of weighing what it's doing to you. By clinging to the belief that fuels your negativity, you aggravate the reaction. Let's use rage as the reaction. The reason people get stuck in angry, hostile, raging behavior has a direct link to core beliefs.

Beliefs That Keep You Mad as Hell

I have a right to act any way I damn well please.

"They" are bad people who deserve my anger.

Getting mad is a healthy way of blowing off steam.

I can't help myself—my emotions run away with me.

Righteous anger is moral, which makes it all right.

Human nature is horrible to begin with.

The person who made me angry is responsible, not me.

I'm not hurting anybody when I get angry.

Anger is an effective way to get what I want and show who's boss.

Each of these beliefs is self-justifying. Each reinforces itself, and the longer you remain stuck on it, the more ingrained such a belief becomes. Deeply ingrained beliefs feel "like me—this is just who I am." But in reality you are hiding from yourself, and rage is doing harm to you. A rage episode is both stressful and inflammatory. But to see reality clearly requires self-awareness. Most people use anger as a weapon, with the intention of attacking others, acting out of self-defense, expressing pent-up frustration, or getting what they want through intimidation. (A minority, chiefly bullies, simply like the feeling of being angry.) These secondary benefits feel so important—or have become so conditioned over time—that the real harm being done is discounted.

Not everyone has a persistent problem with anger management, of course, but all of us tend to normalize our bad behavior. For example, in a family in which the children witness their father abusing their mother, either physically or mentally, this aberrant behavior becomes their version of normal. Even when children from backgrounds of domestic abuse hate the behavior and grow up to renounce it, their risk of turning abusive themselves is much higher than average. They've been too conditioned by growing up in a family in which abuse isn't taboo. In their minds there's a confused imprint: "Daddy hits Mommy" is next to the understanding "Daddy loves Mommy." The contradiction between these two imprints is very hard to resolve when both imprints are part of your childhood training.

Today we are asking you to make choices that bring these old, unconscious imprints to the surface so that they can be examined and healed.

The "Do" Side

Our "Do" recommendations revolve around bringing core beliefs to light and examining them. Your beliefs have tangled roots, and only you can untangle them. Some formative influences are universal, such as family attitudes, upbringing, religion, peer group attitudes, and everything that happens in school. But externals cannot account for why one person is so deeply affected by an experience that rolls off another person's back. We aren't asking you to psychoanalyze yourself, much less to judge yourself for being bad, wrong, inferior, and so on. Our aim is simply to bring core beliefs to light so that you have more freedom of choice in how your life is going. Self-awareness is a healing force. It doesn't always bring immediate healing, but it starts you out on the path.

Once you begin to see why you hold certain negative beliefs, you can retrain the bodymind, and over time, imprints from the past won't have a strong influence on how you think, feel, and behave. The steps of retraining aren't mysterious, and all are under your control. Whenever you feel the grip of a thought that makes you experience tense, angry, guilty, ashamed, or judgmental feelings, go through the following steps:

1. Recognize the negative thought and look at it.

2. Say to the thought, "I don't need you anymore. You can go."

3. If the thought reminds you of bad things in the past, say to yourself, "I am not that person anymore."

4. Sometimes the negative thought is so insistent that it doesn't fade away immediately. Repeat the affirmations just mentioned a few more times. Lie down, take a few deep breaths, and center yourself. (In a work situation, try to find a quiet place and center yourself.)

5. Keep breathing and allow your attention to go wherever it wants to, without resistance. As you do this, ask for relaxation to enter in. Keep going until you feel the tension or discomfort begin to fade.

6. To counter the actual content of the negative thought, replace it with something realistic and optimistic—both need to be present.

For example, if the thought is "I'm helpless. I'll never get through this," the underlying belief is one of victimization and hopelessness to deal with stressful challenges. To retrain yourself, write down all the countering thoughts you can think of.

In this case, the realistic, optimistic thoughts might include the following:

~ *"I'm not really helpless. There are solutions if I look for them."*

~ *"I've survived worse crises before."*

~ *"Feeling helpless is just a feeling, not a reliable way to judge the situation."*

~ *"I don't have to do this all alone. It's okay to ask for help, advice, guidance, and so on."*

~ *"I want to stand up on my own. I welcome this as an opportunity to grow."*

This kind of retraining is absolutely critical if real, lasting change is to occur. Core beliefs are like icebergs showing only the tip above water. In fact, that's what makes the word *core* appropriate. When you encounter someone who persists in self-defeating behavior, something deeper in the psyche is generating it, like a microchip constantly sending out the same signal. In his critically admired memoir *Born to Run*, Bruce Springsteen speaks with unusual candor about the roots of his drive to become a rock star. He grew up with a father who drank heavily, a brooding man who offered no encouragement or support for his son. Springsteen declares that his chief memory of his father is that he sat in the dark in the kitchen drinking and never speaking. Springsteen believes his father hardly spoke a thousand words the entire time he grew up.

This powerful imprint had far-reaching consequences, but they can't be labeled strictly negative. The classic rock song "Born to Run" became an iconic Springsteen hit, but it revealed his own motivation in life, to run away from an emotionally disabled father, to find his own authentic self, to make something out of his talent, and above all to keep running. Added to the emotional mix was a powerful source of love, which came from Springsteen's grandmother. She had lost a very young daughter who ran out into the street and got hit by a car. Her chronic grief never lifted, and when Bruce came along, she found an intense, burning focus for all her maternal love.

Bruce's parents were too poor for one of them to stay home and raise their child, and his care was given over to his grandmother. As a result he became a little prince (or tyrant) enveloped in her all-consuming love. Springsteen looks back upon those years with mixed feelings, knowing that his grandmother's love filled a hole in his heart but also seeing the obsessive and unrealistic side of it. Psyches don't balance positive and negative influences on a scale with one side outweighing the other. Instead, the now familiar cloud of influences develops in an amorphous way that is hard to define or grasp.

For Springsteen the urge to escape was overwhelming, and music became his salvation. But if we turn back to the Harvard study on aging, the point was made that money and fame cannot substitute for close, loving relationships. Springsteen found himself unable to sustain a close relationship, always finding ways to drive girlfriends away. With considerable insight he identifies the underlying dynamic. He had a core belief that he was unlovable; therefore, when anyone got too close, he lashed out, punishing the other person for having the effrontery to dare to love him.

Caught in this psychological knot, Springsteen discovered that he was in a double bind: the thing that would heal him, love, was the very thing he most feared and pushed away. In time, he went through intensive psychotherapy, and he was fortunate to marry a woman who "whipped me into shape," as he puts it. In other words, she loved him enough that she refused to be pushed away. It takes real courage to undo a core belief that you are unlovable. Often what's involved is a formative experience in which the people who are supposed to love you, your parents, are at the same time the source of your deepest hurt. Springsteen was in essence an abandoned child, even though his grandmother's love could compensate for that to some extent. Unfortunately, compensation isn't the same as healing. Today, past sixty-five and still needing to perform as both artistic expression and self-preservation, Springsteen has walked the path of self-awareness, but the depth of his early emotional abuse exacts a price in bouts of severe depression.

So what can we learn from this story, aside from the fact that it came from a very famous public celebrity? To us, the story reinforces the points we've been making about how healing works. The earlier you face your past wounds, the better. Running away and acting out only work in the short run and are obstacles in the long run. Yet with enough self-awareness healing is always possible, beginning with the belief that you want to heal and deserve to.

The "Undo" Side

When it comes to core beliefs, doing and undoing are merged. No one escapes the grip of their deep-held beliefs, and when you seek healing, there are old imprints that must be erased in order for new beliefs to really sink in. Retraining will always involve untraining. To begin the process you need to become unstuck, which is the focus of our "Undo" choices. Harmful beliefs aren't one-dimensional. They create consequences to your cells, perhaps even your epigenome; they are significant for your level of stress and inflammation; they dictate your reflex reactions; and in the end they are invisibly woven into your mood, emotions, and even your view of life.

This is an area in which you need to commit yourself to inner transformation. Mindful techniques and meditation clear the way by providing more detachment and self-awareness. You begin to identify with a level of the mind that doesn't need anger, fear, stress, agitated feelings, and constant drama. Having experienced a state in which these things don't exist, you will naturally begin to question the other state, which countless people are stuck in, in which anger, fear, stress, agitated feelings, and constant drama are so common, they're simply a given. Opening into a higher state of consciousness is a lifelong journey, and core beliefs are just one aspect. But they are very useful as an example of how being attracted to the good is matched by being pushed away from the bad.

What we're asking you to undo is rigidity, closed-mindedness, mindless habit, outworn beliefs, attitudes that bring stress to yourself and others, and us-versus-them thinking. These things only get undone through expanded self-awareness.

To be totally frank, self-awareness has its skeptics and detractors. Society drums in the axiom "What you don't know can't hurt you," despite the dozens of reasons why this isn't true. Inertia keeps us stuck in place; fear of looking at the underside of our behavior promotes a state of denial. With these defenses in full swing, it becomes easy to believe that becoming more aware of our problems will only lead to more hurt. There's no doubt that if healing was about digging up the pain of the past, few people would make the effort. But the whole picture doesn't look like that. Once you start to become aware of a harmful belief, the pain that returns isn't the same as the original pain—this time you can reflect upon it and consciously address it. Pain that has control over you is far worse than pain you can control. On the other side, the experience of being at peace, pain-free, and self-accepting is pleasurable and motivates you to stay on the healing path.

There are endless possibilities for believing in this or that, but one single factor, we believe, binds all negative core beliefs—judging against yourself. Self-judgment is so painful that people will do almost anything to escape the guilt and shame it induces. When Bruce Springsteen discovered deep inside that he was punishing the women in his life for daring to love him, he hit one of the most distressing kinds of self-judgment. Self-judgment isn't about how you act, think, feel, or behave. It goes to the very heart of who you believe you are, your very identity.

What you believe about yourself has both positive and negative effects. If deep down you believe "I must be successful at all costs," you will gain a strong motivation, which is positive. But if you believe success involves ruthless, selfish, and hurtful behavior, your motivation is disfigured by the belief that you are the kind of person who has no choice but to be ruthless, selfish, and hurtful. This is what it means to have your belief control you instead of the other way around. "I'm here to become a success. I don't need anyone to love me" is a defense against facing the issue of whether you are lovable to begin with, and other forms of self-judgment are usually involved, such as "I have to take care of myself because no one else wants to" and "I don't want anyone to see how weak I really am, so I'll always stay on the offensive."

Love is something everyone naturally wants to give and receive, until self-judgment compromises this desire. But it doesn't stand alone. There are four core beliefs that heal self-judgment:

- I am loving and lovable.
- I am worthy.

- I am trusting, safe, and secure.
- I am fulfilled.

You already hold core beliefs in these four areas of love, self-worth, feeling secure, and feeling fulfilled. For most people there's confusion and compromise that blocks the pure feelings that should naturally be there. With self-awareness you can reach clarity by finding a level of the mind where love, self-worth, security, and fulfillment can be experienced directly, without the shadow of a doubt. When you are settled and at peace after meditation, for example, self-judgment is absent. The same is true when you first wake up in the morning or just before falling asleep at night. In all these instances your ego-personality, complete with all the beliefs that reinforce "I, me, and mine," has withdrawn, but your aware-ness hasn't. You are experiencing wakefulness as a steady state. We'll expand on this state on Sunday, the day devoted to evolution (page 253). For now, we just want you to be aware that getting unstuck from self-judgment doesn't have to be a trial; the wakeful state is the easiest, most natural state to be in.

If you look in the mirror, where do you stand on being loved and lov-able, feeling worthy, feeling secure, and experiencing inner fulfillment? Many people would privately concede that there is a lack in these areas, but they don't know what to do. First, realize that no one was born with core beliefs. The issues of love, self-worth, security, and fulfillment evolve as life unfolds. Society offers little reliable guidance, so core beliefs are decided at the level of the private self, which confronts emotions, and the higher self, which provides vision, meaning, and purpose. Emotions pull us up and down, this way and that. The higher self always brings us to our center.

Therefore, the strategy for healing your core beliefs lies with the higher self alone. *Higher* implies many things, but it shouldn't mean something far out of reach. We will soon discuss this more. Here we just want to say that finding love, self-worth, security, and fulfillment is a process. If you enter the process, you will find these things within yourself. They aren't a matter of struggle and strain. Your higher self wants to give you your heart's desire. With this in mind, healing your core beliefs is a matter of connecting with your own true nature. What could be more inspiring?

Non-struggle

DO

Take an allowing attitude.

Approach a situation without resistance.

Act gracefully.

Share a responsibility.

Encourage areas of flow.

UNDO

Stop resisting where you don't need to.

Let someone else have their way.

Help reduce an area of conflict.

Remove obstacles from someone else's path.

Ease competition in favor of cooperation.

"Non-struggle" isn't a familiar term, but we're using it to embrace three things that are familiar: surrender, acceptance, and flow. *Surren-der* is about giving up your attachments, whether to a grievance, which would be a negative attachment, or to a wish that will never come true,

which is a positive attachment. Positive and negative aren't all that important if an attachment keeps you stuck in place. *Acceptance* is about the truth that reality is never wrong. In human life reality is dynamic and moving. Wherever it wants to go, that's the direction that will prevail even if we resist because life seems to be heading in the wrong direction. *Flow* is about approaching life as a smooth, self-directed stream of events.

When surrender, acceptance, and flow merge, you are leading your life without struggle. As a formulation, this sounds very appealing, but society imposes a value system that tilts strongly in another direction. Society teaches us, especially in the West, that surrender is what happens when you are on the losing side of a struggle. Acceptance is a resigned sense that what you want isn't going to happen—you have to settle. Flow is what rivers do, not what is needed to face the hard realities of life.

A much larger belief system lies behind these negative connotations, one that insists that struggle is necessary in order to survive. Struggle has a mythic foundation in the Old Testament, the fabled Fall of Adam and Eve. The Fall came about when Eve persuaded Adam to eat the fruit of the tree of knowledge. Suddenly the first humans knew shame over their nakedness and earned punishment for the sin of disobeying God. The Fall was a catastrophic event—God punished Adam and Eve by throwing them out of Paradise and condemning them to lifelong toil and suffering.

Leaving aside the religious implications, the story of the Fall explains the human condition, and that's where the choice between struggle and non-struggle still rests. Deep down we all hold on to beliefs that tell us how life is—and must be. The last phrase is important, because if life "must be this way," we are powerless to change it. Consider the three indices of happiness used by the Gallup organization to measure well-being (page 154): suffering, struggling, and thriving. Without a doubt there is immense suffering in the world, but that's not the same as saying there *must* be suffering in the world. Not unless your belief system tells you so.

Today we ask you to examine your connection to struggle as something you have to accept. Ironically, people who participate in lifelong struggle have accepted and surrendered in their own way; what they accept and surrender to is a belief that struggle is unavoidable. The opposite worldview would be something like Buddhism, which holds that pain and pleasure are inevitably connected to each other, and therefore the way to rise above suffering is to stop participating in the cycle of pleasure and pain. To do this, a person seeks and finds a level of self-awareness that is eternally the same, eternally peaceful, and undisturbed by the mind's ceaseless activity.

This worldview, which opens the way to non-struggle, also has its own "must": the seeker must be mindful, must turn his back on the pursuit of pleasure, must focus on expanded self-awareness, and must accept that the goal of non-struggle is possible. The reason that most people don't reach the goal isn't mysterious: they find it too difficult to follow the "musts" that this entails. Let's set aside the specific teachings of Buddhism. From an everyday perspective, people want to stop struggling. There's no need for a higher doctrine or teaching—the naked experience of beating your head against the hard realities is motivation enough.

A moment of soul-searching is needed first. Think of an aspect of your personal life in which you find yourself struggling. Here are the main areas where you might look:

Struggling with yourself

Struggling in your relationships

Struggling to improve your life materially

Struggling against the world and external forces

Your struggles, big or little, are likely to fall into these four categories, and if you keep looking, more examples will probably come to mind. A person caught in the throes of addiction or depression is at one extreme of struggling with themselves—the fight is "in here." Another person who is resisting his outbursts of anger or who wants to live up to religious ideals (such as avoiding sin and temptation) experiences the middle ground of self-struggle. Someone who has a high degree of self-acceptance and self-worth experiences smaller struggles, such as trying to maintain a desirable weight or remain young. In short, no life is without areas of struggle, even when someone falls into Gallup's category of thriving.

Because there are so many ways struggle can manifest, people miss the question that matters most: Is any of this struggle actually necessary? Without paying attention to the question, people keep on living as if the answer is yes. They struggle because they feel they have to. To see how this works, consider the following list, which uncovers the psychological attitudes behind everyday struggle.

Why Are You Still Struggling?

I can't see a way out.

I'm in a bad place emotionally (depressed, anxious, helpless).

I feel conflicted and confused inside.

The situation is complicated.

I made bad choices that I'm stuck with—you can't turn back the clock.

It's been this way as long as I can remember.

I'm too scared to fight back.

It's not me—life is hard.

I'm in too deep—I feel overwhelmed.

Someone else is in control of the situation.

I've got no one to turn to.

I deserve this.

These are the most common rationales for getting mired in struggle, not to mention suffering. If you find yourself in a critical situation that severely tests your ability to cope, such as a nasty divorce or declaring bankruptcy, everything on the list might run through your mind at one time or another. Pause for a moment and go back mentally to a difficult time in your life. Can you identify with things on the list that keep you stuck and unable to move? Rationales are powerful because there is an aspect of "must" in them; otherwise, you'd find a way out instead of wasting time and energy on rationalizing why you are stuck.

We're not saying that you or anyone else is to blame for your struggles. Some situations are unavoidable, and external forces are always at play. Getting fired, having to care for a parent with dementia, dealing with a teenager using drugs—life brings countless tests. But you add to your difficulties by attaching "must" to them. Getting to a place of non-struggle is the same as eradicating the "must" from your worldview.

The "Do" Side

Life flows as you choose to let it—that's the central theme of today's "Do" recommendations. The bodymind is designed for unobstructed flow. Information moves everywhere freely; processes are interwoven; the same purpose—to live and thrive—is shared by every cell. When the flow is blocked, the bodymind runs into resistance and obstacles. This is an internal situation—we decide for whatever reason to accept the need for struggle. Once a "must" is in place, it tends to go viral. Your attitude infects others around you, and because your "must" insists on having its way, situations mirror your inner world.

By the same token, if you trust that life can take care of itself, which is the bedrock of every cell in your body, outer reality will begin to conform to your inner world. Optimism, allowing, nonresistance, tolerance, self-acceptance—these can go viral, too. This phenomenon can only be known by testing it. Sociologists already have tested it, after a fashion. One of the largest databases for lifestyle choices is the Framingham Heart Study, which began in 1948 with 5,200 residents of the town of Framingham, Massachusetts. Although the principal aim of the study is cardiovascular health, sifting through the data revealed an inexplicable finding.

A person's overall risk of having a heart attack includes their family background. Someone raised in a household where there is smoking, a sedentary lifestyle, obesity, and so on is more likely to adopt those things in their own life. Extending this fairly obvious connection, if someone belongs to a circle of friends who smoke, lead a sedentary lifestyle, or are obese, the chances of following suit are increased. But the inexplicable part enters with friends of friends. The tendency to make a specific life-style choice like smoking increases outside the circle of people you know. For example, if your parents smoke, you smoke, and your friends smoke, there is an increased risk that the people your parents and friends know will smoke, too, even though you never meet them. In other words, a habit can go viral.

Because there are good habits as well as bad, it's a small step to see that if you grew up in a loving family, which would decrease your risk of heart attack, you will be loving, you will have loving friends, and somehow, *their* friends are more likely to be loving. So the data from the Framingham Heart Study would indicate, even though no one knows what the actual explanation is. Our point is that if you take an attitude of accepting, surrendering, and letting things flow, the effect can go viral. Reality around you will display less struggle and more non-struggle.

To prove that this is possible, you must test it yourself by trying out the "Dos" on our list. If you find yourself in a situation today that cries out for you to step in, intervene, take control, assume full responsibility, tell others how to behave, and so on, recognize this as the perfect opportunity to see if the situation can flow into a good resolution without your intervention. Even if the outcome isn't perfect, you will be surprised at how well non-struggle works. Flow is a real phenomenon, and the more you become convinced, the more you will realize that the "must" behind all your struggles isn't necessary to hold on to.

The "Undo" Side

If flow is a real phenomenon, why don't we see it at work all the time? Because we create internal resistance and roadblocks to control life. We can't really blame ourselves for this desire. It's hardwired in us to do what we can to survive, and in our fast-paced world today most of us are living at a hypersurvival level. Whether it's realistic or not, we want to control the world around us. The "undo" recommendations for today focus on becoming aware of your resistance as it occurs. Specifically, you are stopping the flow whenever you

Create stress for yourself or someone else.

Insist that you are right and others are wrong.

Act judgmental.

Make it your way or the highway.

Refuse to listen to outside voices.

Put someone else down in public.

Impose your own morality.

Today be vigilant about how these behaviors are cropping up at work, in your relationships, or in family life. We all justify our behavior, so it might be easier to observe if other people are acting this way. Then you can reflect on the part you're playing. For example, if something as trivial as arguing over a movie or TV show comes down to "I'm right" versus "No, you're wrong," the tug-of-war needs two people to hold their end of the rope.

Whenever you become aware that you are blocking the flow, stop and get out of the way. This may mean literally walking away or else altering your behavior. In the world's wisdom traditions, reality "out there" mirrors reality "in here." Whether you fully accept that every situation is a self-reflection, you can remove roadblocks and stop resisting, then observe for yourself if the external situation shifts of its own accord.

"Life Can Take Care of Itself"

If the bodymind has evolved to take care of itself in countless exquisite ways, could the same be true everywhere? The question points in a spiritual direction, because it asks if life itself is designed to support human beings. Are we so special? In spiritual traditions both East and West the answer is yes. By teaching that the soul or higher self is real, a long line of sages, saints, and spiritual guides has affirmed some basic truths:

Nothing is random. Every experience fits into a larger scheme.

The larger scheme is embedded in consciousness.

Everyone is connected to the larger scheme, whether they know it or not.

To understand where you belong in the larger scheme, you must expand your awareness.

No matter how you define "the larger scheme," these teachings are totally absent in secular society. Nothing like God's plan, redemption of the soul, karma, or nirvana fits the modern secular model. Two worldviews clash, and the repercussions touch everyday life. In the spiritual worldview human beings are cherished in the universe, which is governed by a cosmic mind; in the secular scientific worldview human beings are a speck in the black void of outer space, existing on the same level as hydrogen atoms or the Milky Way as a product of random chance unfolding after the big bang. Between these two opposing worldviews there is no compromise—the choice is either/or.

This is true in the abstract, but in daily life people fence-sit. How often have you heard the following remarks?

There are no accidents.

Nothing is really a coincidence.

Everything happens for a reason.

Be careful what you wish for.

No good deed goes unpunished.

As you sow, so shall you reap.

You can believe any of these things and also believe that getting rear-ended in traffic is an accident. Our mind inhabits both realities, crossing from one to the other as we please. When someone says, "Everything happens for a reason," the implication is that there is a hidden pattern in everyday events. This hidden pattern offers glimpses of itself, but only glimpses. By now almost everyone knows the term *synchronicity*, which sees a meaning in coincidences. Freud, being a committed atheist and a scientist, had no use for higher powers, the soul, spiritual experiences, or synchronicity. His rebellious acolyte Jung, who invented the term *synchronicity* (defined as two events that are meaningfully connected but have no causal relationship), couldn't persuade him.

As quoted on the website Physics Forums:

The first real crisis in their friendship came in spring 1909, from the following incident. Jung visited Freud in Vienna and asked his opinion on precognition and parapsychology. But Freud was too materialistic and rejected these matters in a way that upset Jung. A strange thing happened then. As Freud was leaving, Jung felt his diaphragm burning and a very loud crack came from the bookcase next to them. When Jung told Freud that this is a perfect example of paranormal phenomenon, he still denied it. Then Jung predicted that in a moment there would be another loud noise. And he was right; a second loud crack came from the bookcase. Freud remained puzzled and this incident raised his mistrust toward Jung.

What really happened that day? The line between synchronicity and the paranormal has been blurry from the beginning, but the larger issue is this: Do our minds affect reality "out there"? People silently answer the question when they believe in things like "Be careful what you wish for." To fully accept that inner and outer reality are connected, you'd commit to beliefs like the following:

God is always listening.

We live in a conscious universe.

The human mind is a reflection of cosmic mind.

No prayer goes unheard.

If you wish hard enough, dreams come true.

So in everyday life, we all have divided allegiances. By seizing on even a faint belief that the world "out there" reflects who you are and what you desire "in here," you can test the truth. In this chapter we've offered non-struggle as true and real. Despite all the struggle you see around you, none of it *must* exist. This can be one of the most profound realizations in your rise to higher consciousness. You can experience non-struggle by following your own personal journey. The division between "in here" and "out there" never existed in the first place. The exquisite way in which the bodymind functions as a whole tells us that, indeed, life can take care of itself. No other proof is needed. It sounds lofty to say that someone is on a journey to higher consciousness, but the truth is more humble. The journey takes us to the state of trust, acceptance, and flow that sustains every cell.

Evolution

Today's recommendations—choose only one.

DO

Be on the lookout for synchronicity (meaningful coincidences).

Change your daily narrative for the better.

Look for a chance to be compassionate.

Openly express love and appreciation.

Be generous of spirit.

UNDO

Resist the voice of fear.

If you find yourself expecting the worst, step away from that expectation and remain neutral.

If you have a negative thought that keeps returning, ask if it is truly serving you or is a relic of the past.

If you feel emotionally upset, find a quiet place to become more calm and centered.

Seek the company of people who inspire and uplift you.

*Sunday is an appropriate day to reflect upon your highest values. Ev-*eryone has aspirations. Everyone wants a life filled with meaning and purpose. The results of these desires take decades to unfold. In the end, however, those who arrive at old age feeling fulfilled will enjoy a higher quality of life than those who look back with regret, frustration, and nostalgia, even if their lifespan in years is the same. We've spent most of this book talking about how negative influences like stress and inflammation build up incrementally over time. But the same is true of personal growth. The soul ripens incrementally, day by day. When this happens, life is a rising arc from birth to death. So how can this vision be turned into reality?

Our whole-system approach to the bodymind has blossomed into a healing lifestyle that will bring benefits for a lifetime. The final step is to take a whole-system approach to life itself. For that to happen, you need an all-encompassing vision. Religion provides just such a vision. Think of the statements a staunch believer can make that apply to all of life, such as the following:

It's all in God's hands.

Faith will carry me through.

God is all-merciful.

As you sow, so shall you reap.

Man proposes, God disposes.

These are all-embracing statements of belief, and if you adhere to them, your entire life will be directed in ways that don't apply to a staunch atheist. Atheism leads to another set of all-embracing statements, such as the following:

The universe is ruled by random events.

Miracles are fictitious.

Religion is an irrational superstition.

Choices need to be based on reason and logic.

It's easy to see that a whole-system approach to life is more common than you might suppose at first glance. Leaving aside religious issues, many people say things like "The family is everything" or "Success is ten percent inspiration and ninety percent perspiration." But is there a similar mindset that applies to healing? Can you stand above day-to-day events and hold on to an all-embracing vision that applies to life itself?

The most successful vision that fits the bill is evolution, a theory that accounts for every life-form, from one-celled micro-organisms and blue-green algae (both billions of years old), to a baby born in a hospital, to you reading this sentence. If you can evolve personally throughout your lifespan, you will have secured an all-embracing vision. Today our recommendations focus on your personal growth/evolution and how to maximize it. To begin with, set aside Darwinian evolution, which is confined to the survival and nonsurvival of species, which means very large groups. Darwinism explains why the saber-toothed tiger emerged from primitive ancestors and finally declined until it became extinct. But Darwinism doesn't tell us anything about a single saber-toothed tiger as an individual.

This is because survival and extinction are ruled by genetic mutations that spread throughout a plant or animal population. If the mutation leads to a survival advantage, it becomes embedded in that species. Human beings long ago escaped this setup. Instead of only the physically strong surviving, we take care of the weak (through health-care plans and retirement payments, for example), and competing for a mate isn't through physical combat. A poet has as much chance to win a loved one's hand as a weight lifter.

There are many arguments over how and why *Homo sapiens* evolved, which we won't enter into here (a whole section of our book *Super Genes* covers the territory). For the purposes of healing, only one point proves critical—the personal evolution of the individual. Evolution for an individual is happening now. We've already offered one validation of this point in epigenetics, which has shown how experiences over a lifetime imprint markers that affect gene activity. Some researchers even hold that epigenetic markers from the mother or father can set their baby's view of how life works (see page 163).

These clues point in the right direction, and so does the evolution of the human brain. Traditionally, the brain is viewed as divided into three parts, from the oldest to the most recent. One can visualize the tripartite brain as being like the staff of an English manor house, which in this case houses the mind. The manor perpetually buzzes with activity, and each of us is the lord or lady overseeing the servants, who correspond to every region of the brain. Downstairs bustles the oldest brain, the reptilian or lower brain, which is nearly half a billion years old. It is organized around survival instincts like fight-or-flight, the urge to mate, and so on, instincts that first emerged in fish and primitive reptiles. Halfway upstairs is the limbic system, which is organized around emotions and bonding. It arose as long ago as 250 million years with the first mammals, who so far as we know are capable of something resembling human emotions (for example, elephants grieve over their dead and porpoises come to the aid of their sick and injured). Somehow the limbic system acquired the ability to remember pleasurable and painful experiences, and from there arose our desire to repeat the pleasurable ones and avoid the painful ones.

At the top landing of the stairs is the most recent brain region, the cortex, where the elite servants wait on the lord and lady of the manor. Everything we think about and decide is managed here. The cortex surrounds the brain like the bark of a tree (*cortex* is Latin for "bark"). When you are deep in thought, your brow furrows, and strangely enough, furrows, crevices, and grooves turned *Homo sapiens* into thinkers. The cortex in rats and mice is smooth. In cats the surface becomes irregular, and grooves start to appear in primates. Evolved species like higher primates and dolphins have deeper, more complicated grooves. But nothing surpasses the biological origami of the cerebral cortex in humans, which is folded into an intricate map that corresponds to the richness of our mental activity. Language, music, and art occur here. (Shakespeare and Mozart are literally groovy!)

We argue that the real you is none of the activities in these brain regions. The real you is the lord and lady of the manor who observe these activities—every feeling, thought, and fancy of the mind. What links the higher brain to personal evolution is quite unique and yet mysterious. It's the ability to be self-aware. Self-awareness spans the immense territory between "Who am I?" and "This is the real me," between self-doubt and self-mastery. Humans experience a bewildering range of self-generated images as we look in the mirror. Seeing ourselves, we can adopt a wide range of psychological profiles, including the following:

Self-satisfied, egotistical, selfish, and blind to our own faults

Self-doubting, humble, altruistic, and keenly aware of our shortcomings

Introverted, reflective, contemplative, and private

Extroverted, aggressive, competitive, and gregarious

These qualities exist in mix-and-match fashion, and for each trait there are polar extremes. So many possibilities exist, in fact, that one could assign a unique profile to everyone on Earth. Without self-awareness we betray our uniqueness and fall into stereotypes and conformity. Habit and conditioning overtake mindfulness. Going along to get along becomes second nature. If these external forces gain the upper hand, a person can live life by rote, existing more or less as a biological robot.

Because we are self-aware, human beings do not simply live: we also watch our lives unfold. It's not possible to get into the nervous system of a humpback whale, giraffe, or panda, but in some way these creatures have their own species of consciousness. It's not simply the physical resemblance that makes tigers and lions members of the same family as a housecat prowling for sparrows on the lawn. They are linked by behavior, and this behavior goes back to how cats perceive the world. They are hunters, prowlers, capable of stealth, patiently crouching before they pounce, and so on.

Today's recommendations revolve around exploring your unique participation in the human species of consciousness. That's a high-flown phrase, we know, but when all is said and done, planet Earth will thrive or decline depending on one thing: whether human consciousness can evolve. If it can, global warming can be stabilized and perhaps reversed. If human consciousness doesn't evolve, inertia will lead us into deeper risk of calamity.

The "Do" Side

To evolve, you need to get into the habit of noticing where a new perspective goes beyond what you are already used to. Today's "Do" recommendations involve just such shifts. Once you free yourself from your accustomed viewpoint, entirely new levels of awareness become possible. At this moment everyone is living out a story in their heads. A good day adds something positive to the story; a bad day slightly undermines the story. The ups and downs of daily life depend on the themes of your story, such as winning versus losing, loving versus hating, leading versus following, and so on.

The themes we live by are well-known and fairly standard, because we absorbed them from our family, friends, and society.

How Your Personal Story Keeps Going
THE THEMES YOU REINFORCE EVERY DAY:

Mindful or mindless	Giving or taking
Optimistic or pessimistic	Supporting or dependent
Winning or losing	Loving or not loving
Thriving or struggling	Attractive or unattractive
Active or passive	Helping or hindering
Doer or thinker	Hungry or satisfied
Loner or gregarious	Seeking or holding tight
Leader or follower	Progress or inertia
Vigilant or relaxed	Confident or timid
Accepting or challenging	Decisive or indecisive

Life is led by reinforcing both positive and negative themes, because they lend structure to a person's story. Without themes, the story would turn shapeless. Yet positive and negative themes share the same flaw. They bind you inside your story. It's better to win than to lose, for example, but if we heed the world's wisdom traditions, winning and losing are opposites that depend on each other. Therefore, winners will always eventually face loss. Optimism will eventually fail. Loving will eventually lead to disappointment. Evolution occurs when you stop identifying with these themes—known as the state of duality—and begin to measure your life in ways that are not dual, not dependent on opposites. What we're talking about are the fundamental qualities of consciousness that lie deeper in our awareness.

The Core Qualities of Consciousness

Intelligent	Evolutionary
Creative	Self-organizing

Self-aware	Knowing
Self-sustaining	Compassionate
Alive	Truthful
Dynamic	Beautiful

Human beings can consciously evolve by discovering that these qualities are real and attainable. That's what the "Do" recommendations set out to demonstrate. If you align yourself with any of these qualities—we've just suggested a handful on the "Do" side—you will be directing your own personal evolution. But this must be more than an ego choice, because ego choices are based on duality. The ego's reason for being truthful instead of lying is that it gains a benefit or avoids a risk. "What's in it for me?" is the basic ego question. The core qualities of consciousness transcend personal identity. They apply to mind itself, the pure essence of being alive and possessing awareness.

Today you can decide to build your story on these primal themes instead of the ones most people accept and live by. Dualism is insecure. What is given can be taken away. The thing you most desire can wind up being a disappointment. Likes change to dislikes and vice versa. Some people have exaggerated, even one-dimensional stories, like "I'm a winner" or "I'm a cockeyed optimist." But one way or another, people base their stories on themes that are constructs to which we attach ourselves.

Today we are asking you to take a higher view, to watch yourself as you keep living the same story. Only then can you make a choice that bases your story on permanent, unshakable values, like acting out of compassion or expressing love and appreciation. To really be transformed, your story must evolve, and your story can't evolve unless your consciousness evolves.

The "Undo" Side

Everyone believes their own story, even when it's divorced from reality. Think of fashion models who are insecure because in their own minds they aren't attractive enough; their self-esteem is rocked by a pimple or the first wrinkle. Think of a professional baseball player on a dismally losing team who still feels like a winner—winning is what got him into the major leagues to begin with. We hold on to our stories for emotional reasons; therefore, the "Undo" recommendations for today are all about getting free of emotional bonds. The bonds that keep us feeling insecure, unsafe, anxious, pessimistic, frustrated, and unfulfilled are blocking our evolution.

A useful concept here is the "emotional body." It includes the ingrained emotions that sustain you the same way your cells support the physical body. In their emotional bodies one person may include feeling loved, safe, secure, and optimistic while another person feels the opposite. If you try to improve your story, the ideal is to base it on the core qualities of consciousness we've just discussed. But this can't happen if your emotional body is wounded. There is simply too great a gap.

Your emotional body can be healed. Undoing the wounds you have suffered from the past is a viable process anyone can undertake. The symptoms are easy to spot—any strong negative thought that repeats itself is a symptom of pain in your emotional body. Let's look at a few of the best and easiest techniques for dispelling negative thoughts from the emotional body.

1. *Catch your negativity early.*

Once you are sunk deep in gloom or anxiety, it's more likely that you will find it hard to lift yourself up. So be on the lookout for the first signs of negativity. As soon as you spot a mood shift toward irritability, anger, frustration, worry, or pessimism, pause immediately. Take a few deep breaths, center yourself. Let the emotion pass, and get yourself somewhere quiet and pleasant, such as going outdoors for a walk.

2. *Avoid external stressors.*

Dark thoughts usually occur under stress, and if you can, you should get away from the stressor, whether it's a negative person, a tense situation at work, or bad news on TV. Dark thoughts set in when they are reinforced, so don't let anyone or anything reinforce your bad mood if you have the choice to avoid it.

3. *Develop a supportive inner dialogue.*

About 75 to 80 percent of people talk to themselves in their heads, and a small minority even hear inner conversations. When the voice in your head starts saying things that induce worry, fear, anger, guilt, shame, or lack of self-esteem, pause for a moment, and say to the voice "That's not me anymore." Repeat until the dark thoughts depart. You might also try "I don't need this anymore. It doesn't serve me."

4. *Keep company with positive, optimistic people.*

We all have friends and family members who are downers. They are pessimistic or complaining; they insist on seeing worst-case scenarios and failure around the next bend. Inertia keeps us from walking away from these people, and sometimes you're stuck in situations you can't escape. But you can cultivate friendships with positive, optimistic people. Sociological studies have shown that you are more likely to adopt positive attitudes and behavior if you keep company with friends who already display them.

5. *Try a "thought-replacement" strategy.*

A technique that lies at the heart of cognitive therapy (an approach that addresses thoughts and beliefs rather than feelings) is to question negative thoughts by asking if they are actually true. For example, if you begin to feel frustrated and think, "What's the use? Things never work out," these thoughts are tested against reality. You say to yourself, "Actually, things sometimes do work out for me. I have succeeded by persevering. This might be one of those situations."

The secret here is to be specific and honest with yourself. When any negative thought arises, challenge that thought's validity. Instead of "No one loves me," you replace that thought with "My mother loves me, and my good friends do, too. I'm not helping myself by exaggeration and self-pity." Once you get accustomed to a thought-replacement approach, you'll be amazed at its effectiveness. Moods follow thoughts, which is why discovering that your bank account is bigger than you thought makes you cheerful, while discovering that your credit card balance is twice as high as you thought makes you feel uneasy.

6. *Develop being centered and detached.*

Being detached can be a positive state; it's not the same as being indifferent or bored. Instead, you are centered inside yourself, which allows you to view situations as a witness, without being swayed or emotionally rattled. Detachment develops naturally through the regular practice of meditation, because once you experience the centered, quiet, unshakable level of your mind, you easily learn how to return there at will.

7. *Get "sticky" emotions to move.*

As we've been saying, negative feelings have a mind-body connection, which you can feel physically. After a bout of getting angry or crying, it takes a while before your body settles down. This is due to various hormones, the stress response, and other biochemicals that do not clear immediately. You can help the clearing process by various means:

Taking deep steady breaths

Lying down and resting

Walking outside

"Toning," the technique of letting spontaneous sounds
arise as they will (low groans, moans, shouts, etc.)

Deep, repeated sighs

Everyone needs to have a set of coping skills, and these are among the most useful and effective. Heavy thoughts don't have to cloud your day. You have good choices for how to lift yourself out of them.

The Highest Evolution

Millions of people have stepped onto the spiritual path over the past few decades. The steady decline in organized religion, which began early in the postwar era, doesn't mean that the present generation is less spiritual. Spirituality is about going beyond the union of body and mind to the union of body, mind, and soul. When people step onto the spiritual path, they want to know how it will change them, how their life will improve, whether the dark side of their inner life will be filled with light, and so on.

We haven't discussed these questions very much in this book for pragmatic reasons. Rudy and Deepak both accept the existence of soul, spirit, higher consciousness, and cosmic mind. But those are contentious terms founded on controversial beliefs. Because they are human constructs, there is no guarantee that any of them are more than constructs. What about transcendence, which is the experience of a domain beyond duality? To be practical, we've left spirituality out of the discussion, but spirituality cannot be divorced from reality. Every experience is known through the bodymind, including the highest spiritual experiences. The person who feels the presence of the divine is doing so through the same nervous system that we all possess. Therefore, a healing lifestyle that brings body and mind together opens a portal to infinite possibilities.

Human evolution, having incorporated survival, emotional bonding, and reason, still has new horizons to cross. The highest state of evolution has only one requirement: self-awareness, which the higher brain already expresses. There are many ways to describe the highest stage of evolution—union with the soul, the state of grace, union with God, salvation, satori, going to Heaven. The most ancient term, dating back thousands of years in India, is *enlightenment*. Yet any terminology begs the question of what it *feels* like to reach this highest state. The peculiarity of the spiritual path is that you don't know where you are going when you start out. (This is why the Indian tradition speaks of the pathless path.) The goal keeps shifting and blurring or even vanishing.

In our view, the unpredictability of the path is inescapable. The self who took the first step isn't the self who arrives at the goal. In everyday life this fact already holds true—the self you had as an infant, kinder-gartner, schoolchild, and adolescent has vanished. So it should not be disturbing if the self you identify with today also morphs into something new by evolving. Despite the burdens of the past, with its old wounds and bad memories, we are designed for renewal at every level of the body-mind. New thoughts and new cells constantly replace the old ones.

For all that, there is a measure of what it's like to reach the highest state of evolution: you feel, once and for all, like yourself. You get to *be*. In a detached yet passionate manner, you can observe your instincts, fears, desires, and random thoughts as they arise and fall in your mind. When you can do this naturally, you no longer get stuck in the mind's ceaseless activity as it churns through thoughts, feelings, decisions, and so on. The true self is masked by this activity, as explained in an ancient Indian par-able. A coach pulled by six horses is driving down the road. From inside the coach a quiet voice whispers, "Stop." The driver is amazed, having never heard this voice. Resentfully, he whips the horses faster. Again the quiet voice from inside the coach whispers, "Stop." The driver feels even greater consternation and lashes the horses harder. But it dawns on him that he has never met the owner of the coach, and the one inside the coach must be the owner. He pulls the reins, and the coach comes to a halt.

In the parable the driver is the ego, the six horses are the five senses and the mind. Only when they come to a stop do they recognize that the soul is the master of all. In meditation one can have the actual experience of quieting the mind so that the true self is encountered. You intuitively know that this is a special experience, although it takes time to grow into full wakefulness. Another metaphor is "the light of awareness." Some people actually see an inner light—most often in meditation, but not necessarily—and it holds an attraction that draws them in. Without this attraction, the true self could never overcome the mental activity that masks it. The ego and the five senses demand your attention. The true self gently entices.

It sounds confusing that the world's wisdom traditions make so much of silent mind. In itself, there is no virtue to silence—psychological observations indicate that around 20 percent of people hear no mental voice in their heads. No one knows why this is so or whether it indicates anything good or bad. Silence only becomes valuable when you investigate what's inside it. With expanded self-awareness, silence blossoms, so to speak. In it are embedded the core qualities of consciousness listed on pages 258–259. Creativity, intelligence, knowingness, and all the rest are your birthright. They cannot be fully suppressed, much less extinguished. Simply by being conscious you possess them, but it takes the act of waking up to notice where they reside—at your source. Unlike every other self that can be described, the true self is pure source, pure awareness, pure being.

Consciousness is using your brain to create the world you are experiencing. Your reality is limited to what you are aware of and what you experience. All humans have evolved to inhabit a species of consciousness that's infinitely rich with possibilities. But the highest evolution is to inhabit the reality that fits you perfectly. This is the ultimate healing, the state of complete wholeness. But what proves that such a possibility even exists? The world's wisdom traditions teach that only the individual can prove it to himself or herself. How? By developing self-awareness into a mindful state known as "witnessing." (In some writings it has also been dubbed "second attention.")

When you witness from a detached position, you no longer try to control the details of your life, no longer worry or struggle. That may sound like a state of total passivity, and it would be if you tried to fabricate the witness. If you really wanted to go to a certain restaurant, only to arrive and find it closed, if you wanted to win a competition only to finish out of first place, or if you are attracted to someone who has no interest in you, you could fabricate the response "I don't care. It's out of my control." This is a forced attitude that contradicts how you actually feel. The true witness lies deep within the mind, at the source. It observes every experience from a place of calm mastery. There is no room for loss or disappointment, for the following reasons:

Every experience is infused with bliss at a subtle level.

You experience wholeness, not the play of light and shadow.

You don't have a personal stake in the world.

The play of consciousness in all its modes holds your complete attention.

In the simplest terms, you're the ringmaster of the circus.

If the witness wasn't a natural state of the mind, none of these things would be true. They'd constitute spiritual fiction or wishful thinking. How can you personally determine if spiritual experiences are real? We have an answer that will solve this age-old question.

The "Pull of the Self"

Spiritual experiences, like any other, are verified by having them. Saints and sages aren't descended from a separate species; they were born with the same nervous system as everyone. The reason they reached higher consciousness isn't magical. Instead, they felt an inner force one can call the pull of the self. Nothing supernatural was involved. From day to day, they chose peace over strife, awareness over denial, love over non-love. These qualities are appealing; they pull at everyone.

But other forces also pull at us. Modern society has so much stress and rush, relieved by endless distractions, that a consciousness-based life-style seems out of joint. Going on a meditation retreat shows a stark contrast to all this hustle and bustle, but when you come home, the pull of everyday life is inescapable.

Look at yourself today. How much time will you expend on the duties and demands of work and family? How tired will rushing around make you feel? How much will you long for a distraction to take your mind off everything? In practical terms, this is what the pull of normal life means. The mind is filled with the noise of constant activity just to keep up with everything. By itself, a meditation session isn't enough to counter the pull away from inner silence and self-awareness.

In the world's wisdom traditions such an obstacle was fully recognized. It really doesn't matter if someone lived in ancient India at the time of the Buddha or today in the middle of a noisy city; restless minds have always existed. The solution has always been managing the pull of the self. When you attune yourself to this inner magnetism, as it were, you can maintain your inspiration to grow and evolve over years, decades, and a lifetime.

The pull of the self means reorienting your attention away from external situations, but that doesn't imply that you ignore the outside world or resist it, either. To ignore is a form of denial; to resist only strengthens the hold of what you are trying to push away. Instead, what we're talking about is a new relationship between two worlds, the one "in here" and the one "out there." Think of this relationship as a sliding scale with two end points.

At one end point the pull of the outer world totally dominates. Life will then have certain inevitable qualities, as follows:

- Feeling unsafe and insecure, constantly vigilant to protect yourself from the next threat from outside.

- A sense of insignificance in the face of titanic natural forces.

- Pressure to protect yourself by conforming to social norms and behavior.

- A constant need for outward pleasures, since only they can stimulate a sense of enjoyment from life.

- Fear of disease, aging, and death.

Since most people don't actually function at this extreme, all of this may sound far removed from daily experience, and yet somewhere along the sliding scale we experience degrees of anxiety and stress and often feel overwhelmed by the insecurity that comes from being very small in a very big, empty universe. The pull of the outside world induces us to put physical reality first, and life becomes a struggle to find security and happiness under the threat that everything could collapse at any moment. There are ways to mask our insecurity, like the rush of thrill-seeking, the hypnosis of entertainment, and the drive to succeed. But by turning to the outside world for these things, we only tighten its grip on our attention.

At the opposite end point the pull of the self is total surrender. Life will then have the qualities of full enlightenment, as follows:

- Being centered and quiet inside is a constant state that cannot be shaken by external circumstances. This leads to a sense of complete security.

- One's own awareness provides the joy and fulfillment that life is meant to bring.

- Change is no longer threatening, because you see yourself as the still point in a turning world. Experience passes through you without altering your state of being.

- You live in the eternal now, which makes aging and death irrelevant—
they have dropped away as part of the illusion of change.

- By living from your source, your true self, you are always in touch
with the source of creativity and renewed possibilities.

- You have no conflicts within yourself or with other people, because
the wholeness of pure consciousness eradicates the play of opposites,
including the play of light versus darkness, good versus evil.

This extreme may sound remote, to the point of being otherworldly,
but any experience that draws your attention in this direction has been
caused by the pull of the self. If you pay attention, there are many moments when you feel safe and secure; life looks beautiful; the mind is
quiet and calm; you feel free of regret and worries; the past brings no
bad memories; you find it easy to accept and appreciate your life and the
people in it; an inner joy bubbles up; or you feel somehow that a higher
presence exists and enfolds you.

Everyone values such experiences without being told to; they are satisfying in themselves. It doesn't matter if the feeling lasts two days or two
minutes—the experience feels timeless. Or to be more precise, you slip
out of time into another place that is simply here and now.

If you want to evolve, meditation and making positive lifestyle choices
are important. But evolution won't truly take hold unless you pay attention to the pull of the self. Human beings aren't robots whose wiring
can be changed simply by plugging our brains into meditation, prayer,
positive thinking, or the influence of wise teachers and mentors. The authors are not discounting those things—they have their valued place in
the world's wisdom traditions. But the context of life is always about the
pull of the outside world, which is noisy and fretful, happy one day and
sad the next, full of pain and pleasure in unpredictable proportion. The
pull of the self is quiet but true, oblivious to the rise and fall of everyday
situations. Finding non-change in the midst of change has long been the
byword in the evolution of consciousness. The pull of the self, which you
can notice every day, is the secret for making non-change a living reality.

Paying attention to reality "in here" is how witnessing develops. The
process is simple and natural. No esoteric teachings are involved. Every
spiritual experience is a glimpse of the true self. First you observe it, then

you embrace it, and finally you become it. The transformation follows an effortless flow, so there is nothing to resist.

For a book on healing to be complete, the true self must be held up as the goal of life. We spoke earlier about the mind's bustling activity being like servants in a great manor house. When those servants are gently dismissed, the lord and lady can enjoy the full splendor of the manor. The outside world is theirs, as is the domain of the mind. There are no more constraints, and the spirit grows bright in the enjoyment of absolute freedom. To quote the famous lines from T. S. Eliot's poem "Little Gidding,"

> *We shall not cease from exploration*
> *And the end of all our exploring*
> *Will be to arrive where we started*
> *And know the place for the first time.*

The place he means is inside us, where we find the essence of who we are and have always been. It is our true self, the healing self.

Alzheimer's Today
and Tomorrow

BY RUDY TANZI

I wanted to end on a strong note of hope, touching upon healing a disease that no amount of self-care or medical science has been able to accomplish up to now. Most people approach aging with a sense of dread, despite all the improvements in our social beliefs about growing old, because of the looming shadow of Alzheimer's disease. As lifespan steadily increases, a person's healthspan—the years spent in good health—is often as much as a decade shorter. The threat of Alzheimer's isn't the only reason for this, because other disorders, principally cancer, are chiefly disorders of old age. But none is as feared as Alzheimer's. A 2012 public opinion poll of more than 1,200 people conducted by the Marist Institute revealed that 44 percent of respondents said that Alzheimer's was their greatest health concern versus 33 percent for cancer. When asked what they fear most about Alzheimer's, 68 percent cited being a burden on their families and loved ones, followed by fear of losing memory of their lives and loved ones (32 percent).

Because my professional life as a research scientist has been devoted to finding the cause and potential cure of Alzheimer's, I'd like to explain the disease in detail. Alzheimer's makes for a fascinating detective story that took a sharp and perhaps decisive turn only recently.

It is hard to imagine a disease worse than Alzheimer's. We spend our entire lives, from womb to tomb, observing, learning, creating, and loving, on a journey that moves from one experience to the next. These experiences shape us as individuals and sculpt our personalities. They define how our friends and loved ones see us as unique people in their lives. The brain's neural networks record our experiences and our reactions to them, as memories. Everything we see, hear, touch, taste, and smell is logically placed into context thanks to a rich tapestry of neural connections and interactions that define who we are. This same tapestry allows us to relate to the world; in fact, every quality of sight, sound, touch, taste, and smell depends upon the brain's ability to convert raw neural data into the image of a three-dimensional world.

But like a merciless vandal, as we age Alzheimer's sneaks in and insidiously begins to tear this neural tapestry apart, thread by thread, until the sufferer no longer recognizes their friends and family, who can only stand by and helplessly watch as their loved one disappears. Alzheimer's is a callous and relentless thief of the mind, brutally ripping away the victim's personhood until all is lost, leaving behind a body and spirit disconnected from the brain that brought them to life. While Alzheimer's patients in the early to middle stages may have a relatively well preserved long-term memory, still recalling details of their wedding day, for example, their short-term memory has been devastated. As sensory information is coming into the brain with every new experience, Alzheimer's patients have difficulties placing the information into context and keeping track of it from minute to minute, or in later stages from second to second.

The result is the following set of symptoms (see the Alzheimer's Association website, www.alz.org, for further elaboration):

1. Memory problems that disrupt activities of daily living, especially short-term memory

2. Challenges in solving problems, such as mathematical calculations in paying bills

3. Difficulty with familiar tasks, such as playing a game or following a favorite recipe

4. Confusion with time or place, such as seasons or months, or how to get to certain places

5. Difficulty with reading, driving, or determining distances

6. Trouble following or joining conversations, and frequent problems finding words

7. Misplacing things and finding them in odd places, like car keys in the fridge

8. Poor judgment or decision making, such as being easily conned by telemarketers

9. Withdrawal from usual activities, such as hobbies or following a local sports team

10. Becoming suspicious, paranoid, or increasingly anxious or fearful about leaving home

In 1906, Dr. Alois Alzheimer, a German psychiatrist and neuropathologist, first described the disease in a fifty-five-year-old woman patient, Auguste Deter. Deter had been admitted to a Bavarian asylum called the Irrenschloss ("Castle of the Insane") suffering from what we now recognize as early-onset Alzheimer's, which strikes before age sixty. In most cases this rare form of the disease (accounting for less than 5 percent of cases) is caused by mutations in three different genes (those encoding for the amyloid precursor protein, presenilin 1, and presenilin 2), all co-discovered in the 1980s and 1990s by my colleagues and me at Massachusetts General Hospital and Harvard Medical School. In fact, these were the first Alzheimer's genes to be discovered; they carry over 250 different gene mutations that virtually guarantee early onset of the disease usually well before sixty years old.

We now know that Deter had a mutation in the presenilin 1 gene, the same that plagued Alice in the popular book and movie *Still Alice*, written by my Harvard classmate, neuroscientist Dr. Lisa Genova. In his journal, Dr. Alzheimer wrote that when he entered her room for the first time, Auguste Deter was sitting on the side of her bed suffering from memory loss and hallucinations, which became obvious in his interview with her. Alzheimer also noted that late at night many residents and staff were awakened by her anguished cries of "Oh God! I have lost myself!" That single description perfectly defines this horrible disease: it robs one of one's self.

Presently Alzheimer's disease is becoming increasingly and alarmingly more prevalent, reaching epidemic proportions in the United States and other developed Western countries. (The epidemic has been nicknamed the "silver tsunami.") There were nearly 5.5 million Alzheimer's patients in this country in 2016. In 2017, Alzheimer's and related dementias will cost our health-care system an estimated $259 billion, of which Medicare and Medicaid will spend an estimated $175 billion. This means that nearly one in every five Medicare dollars is already spent on Alzheimer's patients. By the age of eighty-five, one has a 30 to 40 percent chance of displaying symptoms of Alzheimer's. As 71 million baby boomers head toward high-risk ages, Alzheimer's has the potential to single-handedly collapse the entire health-care system.

As a rule, all of us slow down mentally as we get older. Sometime after age fifty or sixty we might begin to have problems recalling names and words. We may also start to lose track of where we put things or experience "senior moments." But just because our brain begins to slow down, this doesn't mean we need to panic. The deficits of aging are compensated for by becoming wiser, gentler, calmer. People would rest easier if they knew that having senior moments is not necessarily the beginning of Alzheimer's. Misplacing your keys is fine—that's usually just a sign of being distracted or not paying attention. But, if the keys were left in your car with the engine running in the garage after getting home from doing errands and such absent-minded events happen increasingly, there might be reason for concern about the health of your brain.

However, some experts argue that the underlying cause could be the presence of small amounts of brain pathology that begin with virtually everyone after the age of forty. My colleague, Harvard neurologist Kirk Daffner, put it this way: As many of us get older we may have "a little bit of Alzheimer's." It's like having a little bit of arterial plaque around the heart but not necessarily suffering from congestive heart failure.

This may all seem scary, but the good news is that we can handle "a little bit of Alzheimer's" pathology in the brain without incurring dementia. We call this *resilience*, which brings into play the brain's ability to compensate. Dr. David Bennett, an Alzheimer's specialist at Rush University, likens it to "the side streets when there's an accident on the expressway. Everything comes to a dead stop, and so you get off and meander through the side streets. You can still get to your destination." The journey will take longer, but you get there. Bennett also takes note of some people who tolerate the Alzheimer's pathology observed by brain imaging, managing to avoid symptoms of cognitive impairment and dementia. Such people often have "a purpose in life, conscientiousness, social networks, stimulating activities—all these things seem to be protective in terms of how your brain expresses whatever pathology it's accumulating."

Further understanding the basis of brain resilience to Alzheimer's despite the presence of Alzheimer's damage requires some understanding of the exact pathology that defines this disease. The trilogy of Alzheimer's pathology includes the following:

1. *Senile plaques*, which are large clumps of sticky material called beta-amyloid that deposit around nerve cells in the brain.

2. *Tangles*, which are twisted filaments that form inside of nerve cells and kill them.

3. *Neuroinflammation*, a response of the brain's immune system to the plaques, tangles, and dying nerve cells. While intended to help as part of the immune system's healing response, this inflammation winds up killing many more nerve cells by "friendly fire."

For decades we didn't know how these three pathologies link to one another, which causes which, or which comes first. This mystery was largely due to early attempts to re-create Alzheimer's disease pathology and symptoms in mice. Researchers took a human gene mutation that causes early-onset Alzheimer's to run in families and inserted it into the mice's genome. The mice got brain plaque but no tangles. This led to twenty years of fiery debate as to whether plaques cause tangles. All those first Alzheimer's genes discovered by me and others indicated that Alzheimer's started with plaques leading to tangles. Yet this could not be proved in mouse models of the disease.

The debate raged on. Did amyloid plaque actually cause Alzheimer's? The familial Alzheimer's genes all pointed to yes, while the mouse studies said no. The implications for treating and preventing the disease were enormous. I previously wrote about this in my 2001 book, *Decoding Darkness: The Search for the Genetic Causes of Alzheimer's Disease*. At the time, the argument was anything but resolved. Since then, much more has been learned. I held that we cannot just trust results from Alzheimer's mouse models. Humans are not 150-pound mice! Then, in 2014, my Harvard colleague Doo Yeon Kim and I decided to settle the issue once and for all. We invented what a *New York Times* headline would dub "Alzheimer's-in-a-Dish," which involved working out a stem-cell technology to grow mini–human brain organoids (an artificially grown mass of cells or tissues) in mini–human brain dishes. Beforehand we placed early-onset Alzheimer's gene mutations into the artificial brain tissue. Miraculously, the mini-brains in the dish formed actual senile plaques for the first time ever, and in only six weeks. More important with regard to the ongoing debate, two weeks after the plaques formed, the human nerve cells were filled with toxic tangles. When Doo and I treated the brains with drugs that stopped the plaques, they also stopped the tangles.

When our study was published in the prestigious scientific journal *Nature*, no one in the field disagreed. The debate was over. Senile plaques do cause the toxic tangles that go on to kill nerve cells. The *New York Times* called the breakthrough "groundbreaking" and "game-changing." Now the discovery of Alzheimer's drugs could be achieved ten times faster and ten times cheaper than in mice. (For this discovery Dr. Kim and I were honored in 2015 with the nation's highest award for innovation and invention, the Smithsonian American Ingenuity Award, and I found myself on *Time* magazine's list of the "100 Most Influential People in the World" for 2015.)

So, returning to a critical question, what makes a person resilient to Alzheimer's? One factor is referred to as "cognitive reserve," which we touched upon at the beginning of this book (page 2). The greater the amount of knowledge one has amassed and learned, for example through higher education, the greater the number of synapses in one's brain. Since the degree of dementia in Alzheimer's patients correlates most closely with loss of synapses, the more synapses you have, the more you can lose before problems set in. Thus, continuing to learn new things is very important as we age. When planning for your retirement, think just as much about your cognitive reserve as about your financial reserve.

Perhaps the most critical information on the nature of resilience has come from some individuals aged eighty to one hundred who die with no cognitive issues, yet who upon autopsy reveal Alzheimer's levels of plaques and tangles. What do these fortunate folks all have in common? In each of these resilient brains there's no evidence of neuroinflammation. Despite abundant plaques, tangles, and nerve-cell death, the brain's immune systems didn't react with the inflammation response. The result was no Alzheimer's disease. In 2008, we discovered a new Alzheimer's gene known by the symbol CD33, which encodes a protein called "siglec-3" on the surface of certain types of immune cells. My colleague Ana Griciuc and I later figured out that this gene is the switch that turns on neuroinflammation. We then found mutations in this gene that could either increase or decrease risk for Alzheimer's by causing more or less neuroinflammation in response to the plaques and tangles that show up in the brain, usually after age forty.

As a result of these studies, many pharmaceutical companies are now developing drug therapies aimed at these genes to curb neuroinflammation. Such a drug would be not only useful for treating Alzheimer's but also helpful for other neurological diseases such as Parkinson's or stroke.

When we put all of this information together, we've validated that the plaques are like a match (head injury can also be the match in other forms of dementia, for example, chronic traumatic encephalopathy), while the tangles and nerve cells they kill are the brush fire spreading throughout the learning and memory areas of the brain. But once neuroinflammation kicks in, it's like a major forest fire, and this is when the symptoms of catastrophic cognitive decline and dementia set in.

Armed with this knowledge, we now realize that we must stop the amyloid plaques first. Brain imaging studies reveal that plaques form ten to twenty years before symptoms of dementia arise. This largely explains why so many clinical trials targeting plaques had failed. They were used on patients who already displayed symptoms, which was at least ten years too late. It's akin to someone being diagnosed with congestive heart failure after suffering a heart attack and then deciding to lower their cholesterol levels. Cholesterol would have had to be addressed a decade earlier. Today, antiplaque therapies are being tried in early, very mild cases of Alzheimer's and even in presymptomatic individuals who have abundant plaques in their brain just starting to initiate the disease process.

I've warned that in these treatments we shouldn't aim at totally wiping out the amyloid plaques. My Australian colleague Rob Moir and

I, with financial support from the Cure Alzheimer's Fund, have discovered that the sticky amyloid plaques actually help protect the brain from viral and other infections. In fact, viruses, bacteria, and yeast can rapidly seed the formation of plaques. This has suggested a new theory about the cause of Alzheimer's, in which the plaques are formed in response to infectious microbes as a natural way to protect the brain.

What does the new theory mean for preventing and treating Alzheimer's? Someday, very early on in life, we may be able to target infections that drive amyloid plaque deposition in the brain. Potentially we could use brain imaging and perhaps blood tests to detect when amyloid plaques are accumulating at alarmingly high levels and then target those plaques with anti-amyloid drugs. Such drugs are currently being tested by pharmaceutical companies and are also being developed in labs like mine at Massachusetts General Hospital in Boston.

Around the same time that we stop plaque buildup in the brain, ten to fifteen years prior to symptoms, it would be best to also stop the tangles from forming and spreading in response to the plaques. It's all about treating the right patient with the right drug, at the right time. For patients already suffering with cognitive symptoms and dementia, neuroinflammation must be curbed. It's too late to only target plaques and tangles.

Until these drugs come on line, what can we do now in our own daily lives to reduce the risk for Alzheimer's as we get older? The following recommendations have been shown to have the most useful effects on risk reduction—you'll recognize them from our general advice for a healing lifestyle, although they are more specific here:

- **Eat a Mediterranean diet.** This is a diet rich in fruits, nuts, vegetables, olive oil, minimal or no red meat, and alternative sources of protein (e.g., fish or, if you're vegetarian, like me, legumes, tofu, and mycoprotein from mushrooms).

- **Get seven to eight hours of sleep per night.** It is during the deepest stage of sleep (delta or slow-wave) following dreams (REM sleep) that the brain clears itself of debris like amyloid plaques. This is also when short-term memories are consolidated into long-term memories.

- **Exercise daily.** Aim for 8,000 to 10,000 steps per day if you have an electronic measuring device. Or take a brisk walk for an hour every day. During exercise, amyloid plaques are dissolved in the brain, neuroinflammation is turned down, and even new nerve stem cells are born in the area of the brain most affected by Alzheimer's, the hippocampus, which is responsible for short-term memory.

- **Reduce stress.** Managing stress with meditation and other techniques protects the brain from harmful neurochemicals like cortisol. In a clinical trial of meditation, we also showed changes in gene expression that favor removal of amyloid from the brain and that lower inflammation. It's also worth noting that as people get older, finding that they can't recall names and words as well, they often become increasingly stressed out, especially if they worry about the beginnings of Alzheimer's. Ironically, this stress can lead to cortisol production in the brain that kills nerve cells, perhaps increasing risk for Alzheimer's.

- **Learn new things.** Learning new things forces you to make new synapses in the brain, enhancing your cognitive reserve. Growing older should include challenges like learning how to play a musical instrument or taking foreign-language lessons, but also small things like brushing your teeth with the opposite hand, taking a new commute route, or simply watching a documentary or attending a lecture. Because all learning is based on associating new information with what you already know, you not only make new synapses but reinforce the ones you already have. Moreover, this leads to new neural pathways for gaining access to information recorded by specific synapses and existing neural pathways. It's worth mentioning that crossword puzzles and brain games do not serve the same purpose as learning new things.

- **Stay socially engaged.** Loneliness has been confirmed as a risk factor for Alzheimer's. Social engagement and participating in positive, supporting social networks have been shown to be protective against a higher risk for Alzheimer's disease.

Some Optimistic
Thoughts About Cancer

Cancer is viewed as a unique kind of threat because of the dread it inspires, but a healing lifestyle is just as pertinent as with heart disease or obesity. Compared with those two conditions, it's hard to get people to feel optimistic about cancer. Fear is a powerful force, all the more so when it contains irrationality. To most people's surprise, cancer is actually moving steadily into the realm of hope and optimism.

After the federal government declared its "war on cancer" in 1971, only to have hopes for a cure fade away, the public has found itself riding an emotional roller coaster. There is still a widespread perception, despite the steady drumbeat that "we're getting closer every day," that progress hasn't been made.

This is a massive misperception that reflects the lingering power of fear. In its 2017 report on cancer rates, the American Cancer Society reported that overall cancer deaths declined by 25 percent between a peak in 1991 and 2014, the latest date for statistics. The reasons for this decline, however, are not related to an overall cure. That goal has been abandoned over the decades once it was discovered that cancer behaves not like one disease but like many. The recent decline has come in bits and pieces. To quote the American Cancer Society's website, "During the most recent decade of available data, the rate of new cancer diagnoses decreased by about 2 percent per year in men and stayed about the same in women. The cancer death rate declined by about 1.5 percent annually in both men and women."

Projecting forward, the bottom-line statistic for 2017 was that roughly 1.7 million new cancer diagnoses would be made and 600,000 deaths would be attributed to cancer. In simplest terms, only 1 out of 3 patients would eventually die from their diagnosis. That's a good baseline for optimism.

For a long time patients feared cancer treatments as much as the disease itself. In the early days of modern cancer therapy, the basic fact that oncology clung to was that cancer cells multiply much faster than normal cells. Therefore, the application of drugs that were toxic to the entire body would hit cancer cells harder. (One of the first forms of chemotherapy was deadly mustard gas, notorious from World War I.) By this logic, if you wanted to kill every scrap of malignancy, it was justified to put patients through severe suffering in an attempt to kill the cancer even if it meant that a certain proportion of patients would be killed first. Today's therapies are much more precisely targeted and safer. More important, they proceed with a new logic in mind, aiming at the genetic foundation of the disease.

Just as important, however, was a dramatic change in attitude. Consider a 2015 article in *The Lancet* that begins with a sentence that would have shocked and baffled the cancer field a generation ago: "The nature of cancer control is changing, with an increasing emphasis, fueled by public and political demand, on prevention, early diagnosis, and patient experience during and after treatment." If you unpack this sentence, it says several important things:

- Prevention is beginning to spearhead the overall approach doctors will take to cancer in the future.

- Cancer is a controllable disease, not always calling for drastic treatment, particularly in older patients with slow-growing cancers such as prostate cancer in its early stages.

- The public fear over cancer is being paid attention to. There are promises of less arduous treatments, and a handful exist.

Looking Deeper

This new attitude toward cancer is a very good sign, but caution is still warranted. Official progress proceeds by small increments. The typical trial for a new cancer drug helps only 3 to 5 percent of participants. And historically the promises about lowering cancer deaths stalled. The toll taken by the disease is known through two measures: first, the number of people diagnosed with cancer each year, and second, the age at which they die. It's the second number that most people overlook. They think in terms of five-year survival, the most common statistic for remissions, which has limited validity.

Early detection is a great boon that former generations didn't have. But it can also increase the survival rate artificially. A woman diagnosed with breast cancer in the 1930s would most likely have been in a more advanced stage of the disease than a woman diagnosed today. Let's say in the 1930s that a woman's doctor detected a suspicious lump when she was fifty-five and she died at fifty-seven after unsuccessful treatment. (Back then, a radical mastectomy was the only viable course of action in this country, since chemotherapy and radiation were still in the future.)

Today, anomalous or malignant breast cells can be detected much earlier, often at stage 1 of the disease if not before. Then it would be typical for the diagnosis to come when the woman is forty-eight, for example, instead of fifty-five. She could survive nine years, which would put her in the five-year survival category, and yet still die at fifty-seven—a different route leading to the same outcome.

This is why the adjusted age of mortality—the average age at which people die who have been diagnosed with cancer—is the key figure. That age needs to increase if we are to claim a real advance in cancer survival. For decades it didn't go up. If you look at the broad picture, cancer deaths have decreased, although not enough, because of interlocking factors:

- Early detection is a boon but can also go too far. The standard test relied upon for detecting prostate cancer, the PSA blood test, led to overtreatment in a cancer known to progress for years or decades without becoming lethal. It was eventually decided that the risk of harming patients through surgery and radiation was actually greater than the projected lives being saved through regular PSA testing (along with false positives from the test).

- A steady decline in smoking has cut into the rates for lung cancer.

- Targeted treatments have become more effective.

- Fewer patients are dying from massive assaults with chemotherapy and radiation than in the past.

- Genetic scanning has made possible new drugs that specifically target the genetic source of cancer, but to date these drugs have been hugely expensive (tens of thousands of dollars per treatment), and few cancers are linked to a single genetic mistake. One exception is a specific form of childhood leukemia, once nearly always fatal but now enjoying a recovery rate over 90 percent (with the serious caveat that the recovered patients run into serious health problems in their twenties).

Yet the main reason for optimism has switched from treatment to prevention. This is a turnaround unforeseen even a decade ago, when hope was overwhelmingly placed on more funding for basic research and new drug treatments. It is now generally agreed that up to 50 percent of cancer cases are preventable using already existing knowledge. Everyday lifestyle choices are the main thrust of cancer prevention, which includes not smoking; eating a natural whole-foods diet; avoiding carcinogens in your food, air, and water; taking half an aspirin per day; and wearing sunscreen.

Most people already knew about taking aspirin to reduce the risk of heart attacks and strokes, and so the benefit for cancer is an add-on, not a panacea. In-depth data gathered from a thirty-year study following 130,000 people found that those who regularly took at least two adult aspirin a week had a decrease in gastrointestinal cancer of 20 percent and in colorectal cancer of 25 percent. (Other studies have corroborated the usefulness of aspirin as a cancer prevention measure, but also to decrease the risk of metastases after a tumor has appeared.)

The reason aspirin is effective in these cancers seems to be its anti-inflammatory effect. One indirect proof of how damaging inflammation is was staring us in the face, when you think of what aspirin is good for:

cold symptoms, pain, and as a preventive of heart attacks. All are linked to its anti-inflammatory action.

The prevention measures relating to sunscreen and not smoking are targeted specifically at skin cancer and lung cancer. But the best news is that positive lifestyle choices which apply generally, such as maintaining a good weight, avoiding alcohol or keeping consumption to a minimum, and leading an active life, are beneficial for cancer—in other words, a healing lifestyle is a broad-spectrum approach. There is nothing you need to do to add an extra layer of cancer protection, because so far as the most up-to-date studies show, no such extra protection exists.

This will come as a disappointment to anyone who tries to lower their anxiety over cancer by resorting to specific supplements, so-called cancer diets, and magical foods that supposedly prevent the disease. One new trend, however, is to link the early formation of cancer with chronic inflammation. To the best of our knowledge, the diet we offer in Part Two of this book comes as close as possible to being an anti-cancer diet as well.

Managing Cancer, Before and After

Finally, there is optimism over cancer being a manageable disease. This is a major change in attitude that is slowly seeping into the medical community. Cancer has always been a desperate "do something, anything" proposition for oncologists as well as patients. The image of an insidious enemy attacking the body from within has motivated immediate, often drastic action. But being a multifaceted disease, not all cancers are created alike. Some are slow-growing, for example. If you consult the five-year survival rate for the seven types of brain tumor, for example, they range from 17 percent for glioblastoma, a deadly aggressive form, to 92 percent for meningiomas, which tend to be benign and so slow-growing that the brain can often adapt to their presence. (Thyroid and bladder cancers also fall into the range of slow, manageable cancers.)

How to manage a cancer depends on the oncologist, and they vary widely in their approach to immediacy of treatment. Consulting more than one is advisable, and it's important to question them about their attitude toward manageability. In any event, there are many factors that affect cancer rates and recovery. Your risk lowers if you are young, white, well-off, and get early detection. You are at higher risk if you are non-white, older, poor, and get delayed detection. (For example, the quoted survival rates for brain tumors applied to the age group twenty to forty-four. For patients fifty-five to sixty-four, the rates worsen, going down to 4 percent for glioblastoma and 67 percent for meningiomas.)

This brings us to an issue that seems self-contradictory, managing cancer before you are ever diagnosed. If you take vitamin C or zinc to ward off a winter cold, you are practicing prevention; it would seem strange to say that you are managing a cold when you don't even have one. But with cancer, the known measures for prevention don't tell the whole story. There is an X factor to contend with, and this X factor must be managed, year in and year out.

We are referring to self-induced stress and fear. Modern society is inundated with medical stress, thanks to constant repetition of stories about risks, studies, tragic deaths, and miraculous recoveries. None of this is more stressful than the news about cancer. Stress cannot be prevented when it is so pervasive, all the worse because you never know when cancer will strike close to home among friends and family. The simplest advice is this: Stress management is cancer management. That holds true for both healthy people, patients who have just been diagnosed, and cancer survivors.

It has become standard post-treatment advice to tell anyone recovering from cancer to seek loving support from family and friends, to which outside support groups should be added. Cancer is an isolating disease. The side effects of chemo and radiation, particularly hair loss and the wasting of muscle tissue, tend to make people want to be alone even more. (The present generation of cancer patients is fortunate that the disease isn't greeted with nearly the same fear it created in the past.)

The reason why managing the emotional stress of cancer is effective remains vague—that's why we refer to it as the X factor. But we strongly suspect that the answer will be epigenetic. As explained on page 163, epigenetics deals with the change to DNA made by everyday experiences. The stronger the experience, the more marks are likely to be imprinted on a person's epigenome, leading to changes in gene activity, since the epigenome, which wraps around DNA like a protective sheath, is believed to be the main switch for gene activity.

Proving that bad experiences may influence the development of cancer in its early stages is fraught with danger, and may increase stress rather than alleviate it. But there is no danger in associating positive experiences with reduced stress. Besides boosting survival rates, pursuing stress management—and specifically working on your underlying fear of cancer—is important long before any sign of disease appears. By acquainting yourself with the optimistic news about cancer, you can take a big step to reducing your level of anxiety. Removing the irrational aspect of our attitude toward this disease might result in the turning point everyone has desired for so long.

Acknowledgments

When a book is being created, it needs as much parenting as editing, so we are grateful for the considerate understanding and nurturing of our editor, Gary Jansen. Also many thanks to others at Harmony Books who constituted and managed the working team: Diana Baroni, Vice President and Editorial Director; Tammy Blake, Vice President and Director of Publicity; Juliana Horbachevsky, Senior Publicist; Christina Foxley, Associate Marketing Director; Estefania Ospina, Associate Marketer; Jenny Carrow, our jacket designer; Elina Nudelman, our book's designer; Norman Watkins, Senior Production Manager; and Patricia Shaw, Senior Production Editor.

More than ever, authors owe gratitude to the publishing executives who are willing to take a chance in these precarious times for publishing. We want to especially thank Maya Mavjee, President and Publisher of the Crown Publishing Group, and Aaron Wehner, Senior Vice President and Publisher of Harmony Books.

From Deepak: I remain indebted to a fantastic team at the Chopra Executive Office, whose tireless efforts make everything possible from day to day and year to year—Carolyn Rangel, Felicia Rangel, and Gabriela Rangel. All of you have a special place in my heart. Sara Harvey and the

Chopra Center staff make special contributions with loving enthusiasm. Thank you for everything. More thanks also goes to Poonacha Machaiah, cofounder of Jiyo, for supporting and promoting a wide range of projects, including this book. As always, my family remains at the center of my world and is cherished all the more as it expands: Rita, Mallika, Sumant, Gotham, Candice, Krishan, Tara, Leela, and Geeta.

From Rudy: I would like to thank my dearest wife, Dora, and the best daughter ever, Lyla, who serve as my personal healers every day with their unconditional love and support. I am also grateful to my mom for teaching me the importance of always striving to maintain a kind, compassionate, and positive outlook in life, the keys to healing at all levels. I would like to thank Susanna Cortese for her invaluable help during the preparation of this book in keeping my research operations going strong. Finally, I would like to thank the Kadavul Temple in Kauai, for the inspiration that came regarding *The Healing Self* after a particularly wonderful meditation.

About the Authors

Deepak Chopra, M.D., F.A.C.P., founder of the Chopra Foundation and cofounder of the Chopra Center for Wellbeing, is a world-renowned pioneer in integrative medicine and personal transformation. He is the author of more than 85 books translated into over 43 languages, including numerous *New York Times* bestsellers. Two of his books, *Ageless Body, Timeless Mind* (1993) and *The Seven Spiritual Laws of Success* (1995), have been recognized on *The Books of the Century* "Bestsellers List." He serves as an Adjunct Professor at Kellogg School of Management at Northwestern University; Adjunct Professor at Columbia Business School, Columbia University; Assistant Clinical Professor in the Family and Preventive Medicine Department at the University of California, San Diego; on the Health Sciences faculty at Walt Disney Imagineering; and as Senior Scientist with the Gallup Organization. *Time* magazine has described Dr. Chopra as "one of the top 100 heroes and icons of the century" and credits him as "the poet-prophet of alternative medicine." The *WorldPost* and *The Huffington Post* global Internet survey ranked Dr. Chopra #40 of the most influential thinkers in the world and "#1 in medicine."

Rudolph E. Tanzi, Ph.D., is a professor of neurology and holder of the Joseph P. and Rose F. Kennedy Endowed Chair in Neurology at Harvard University. He serves as the vice-chair of neurology and director of the Genetics and Aging Research Unit at Massachusetts General Hospital. Dr. Tanzi is a pioneer in studies aimed at identifying genes for neurological disease. He co-discovered all three genes that cause early onset familial Alzheimer's disease (AD), including the first AD gene, and currently spearheads the Alzheimer's Genome Project. He is also developing new therapies for treating and preventing AD based on his genetic discoveries. Dr. Tanzi was named to *Time* magazine's "100 Most Influential People" for 2015 and to the list of Harvard "100 Most Influential Harvard Alumni." He has also received the highly prestigious Smithsonian American Ingenuity Award for his pioneering studies of Alzheimer's disease. He is the coauthor of the *New York Times* bestseller *Super Brain* with Dr. Deepak Chopra, has professionally played keyboards with Joe Perry and Aerosmith, and is the host of *Super Brain* on public television.

ALSO BY THE BESTSELLING AUTHORS
DEEPAK CHOPRA, M.D., AND RUDOLPH E. TANZI, PH.D.

AS SEEN ON PUBLIC TELEVISION

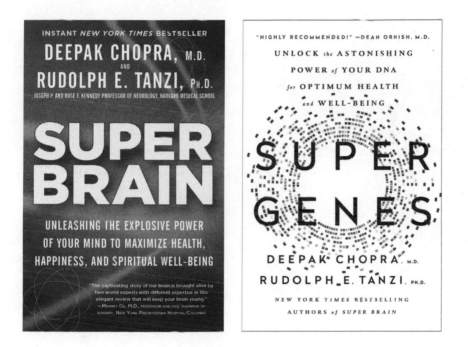

HARMONY

BOOKS · NEW YORK

Available everywhere books are sold.